Non-Prototypical Reduplication

Studia Typologica

Beihefte / Supplements
STUF – Sprachtypologie und Universalienforschung
Language Typology and Universals

Volume 22

Universität Bremen

Non-Prototypical Reduplication

Edited by
Aina Urdze

DE GRUYTER
MOUTON

ISBN 978-3-11-059700-4
e-ISBN (PDF) 978-3-11-059932-9
e-ISBN (EPUB) 978-3-11-059713-4
ISSN 1617-2957

Library of Congress Control Number: 2018937499

Bibliographic information published by the Deutsche Nationalbibliothek
The Deutsche Nationalbibliothek lists this publication in the Deutsche Nationalbibliografie; detailed bibliographic data are available on the Internet at http://dnb.dnb.de.

Cover image: Alpha-C/iStock/Thinkstock
Printing and binding: CPI books GmbH, Leck

www.degruyter.com

Contents

Preface

Reduplication is an incessantly discussed topic in the field of linguistic typology, it involves not only morphology but also syntax and phonology. This volume aims to explore the boundaries of the grammatical phenomenon known as reduplication from an outside angle, i.e. by looking into non-prototypical cases which challenge the formal and functional criteria for reduplication proper. The articles collected here are a selection of contributions which were presented on a workshop held at the University of Bremen in spring 2016. During those two days, 8th and 9th April, fifteen talks on various instances of non-prototypical reduplication were given, both by versed scholars and by graduates. The Bremen workshop was set up with the intent to continue the discussion about grammatical reduplication among linguistic typologists and colleagues with expert knowledge of certain languages. This discussion was initiated at a workshop on total reduplication in Brussels back in 2012. Since that time, a lot of research has been going on and the linguistic circle in Bremen has accomplished several studies – papers, BA and Masters Theses, a dissertation – on this topic.

All these works on reduplication have in common the need to narrow down the scope of research and try to find adequate definitions for their research subject. The extensive study on total reduplication by Stolz et al. (2011) uses the prototype approach in order to cope with the various phenomena which belong (or do not belong) to the realm of grammatical reduplication. Generally, the research community agrees on certain formal and functional criteria which are essential for "proper" reduplication. Formal criteria include completeness and exactness of duplication, as well as contiguity of the two elements. Functional criteria demand that the reduplicative construction should have a meaning on its own which clearly differs from that of the base word. Sometimes the issue of productivity is raised; this calls for systematicity of reduplication and rules out isolated cases.

The idea for the Bremen workshop had its beginnings in this attempt to narrow down the research subject. Instead of leaving aside the instances which do not count as reduplication, we felt the need to deal with the phenomena which deviate far from the prototype and tend to the opposite pole – hence the title of the workshop and this volume: "non-prototypical reduplication". It seems an obvious conclusion to draw, that since there are many instances of proper reduplication throughout the world's languages (Rubino 2013 provides us with a reasonable albeit not flawless overview) there ought to be many cases as well which fall short of the definitory criteria for reduplication proper. It is possible indeed to identify "typical" candidates for non-prototypical reduplication; these candidates are also widespread in the world's languages.

Among the phenomena not (exactly) fulfilling the necessary criteria for reduplication one of the most prominent are echo-word constructions which Stolz (2008) considers a special case of total reduplication. Similar cases can be found where a fixed segment is added to partial reduplication, too. While such echo-phenomena fail to meet the formal criteria, there are also reduplicative constructions which do not meet the functional-semantic criteria in the sense that in these cases reduplication does not result in a new or additional meaning which would be clearly distinguishable from the meaning of the base. Reduplication for purely pragmatic reasons adding emphasis in spontaneous speech matches this kind of non-reduplication. Another instance of functionally-semantically insufficient reduplication is at hand when the base of the reduplicative construction is not/no longer available in a language: for obvious reasons a difference in meaning between copy and (missing) base cannot be determined, either. In the following the different types of possible candidates for non-prototypical reduplication shall be shortly described and exemplified.

– Non-prototypical total reduplication due to formal aspects: echo-word constructions
  Eastern Yiddish uses a reduplicative construction with a fixed sound sequence /šm/ which is either replacing the onset of the first syllable of the copy or added to it. In this way, based on the noun *dikduk* 'grammar' the echo-word construction *dikduk-shmikduk* is formed. The second element *shmikduk* – the echo-word – is a non-word in Eastern Yiddish; neither does it occur on its own nor does it bear any meaning. However, the echo-word construction as a whole has a distinctive meaning, derived from the base word and adding a sense of generic plurality which may include a deprecative shade – the meaning of *dikduk-shmikduk* could be translated as 'grammar and such' (Southern 2005: 21). The formation of echo-words with an initial /šm/ and their use in constructions with generic and deprecative meaning is quite productive and has even spread into Russian and American English.
– Non-prototypical partial reduplication due to formal aspects: intensified adjectives
  Many Turkic languages use a certain kind of non-prototypical partial reduplication in order to intensify the meaning of quality adjectives (very often colour terms). A part of the first syllable (onset and nucleus) of the base word is reduplicated and a specific coda is added, this reduplicated and altered syllable then is positioned left to the base. While other Turkic languages have a set of several consonants that can be added as coda, Turkmen

employs only /p/ in this construction, e.g. *гараңкы*/garaŋkı 'dark' ~ *гап-гараңкы/gap-garaŋkı* 'pitch dark' or *яшыл/ya:šıl* 'green' ~ *яп-яшыл/yap-ya:šıl* 'bright green' (Clark 1998: 510).

– Non-prototypical total reduplication due to functional aspects: pragmatic iteration
  In languages throughout the world one can find instances which look like proper total reduplication but which represent pragmatically motivated repetition. It is often difficult to determine whether a given instance of "double occurrence" belongs to the grammatical or pragmatic sphere, since in the latter case the repetition expresses insistence which is associated with intensification (Stolz et al. 2011: 137ff.). Thus, the repeated adverb in German *das ist viel viel besser* 'that's much much better' signals the emphasis on the speech act as a whole rather than a modified (i.e. intensified) meaning of the reduplicative construction compared to the base word.
– Non-prototypical total reduplication due to functional aspects: non-existent base
  With ideophones and other parts of the expressive vocabulary there can often be found reduplicative structures which do not seem to denote a difference in meaning compared to the single occurrence; these might be reduplications due to pragmatic reasons – see previous bullet – or just lexical variations.[1] However, quite often the base word does not occur unreduplicated, in this sense there is no base at all. This is often the case with Japanese ideophones, e.g. the expression *zukizuki* 'throbbing [experience of pain]' is only used in its (seemingly) complex form and never as merely **zuki* (Otsuka 2016: 33).
– Non-prototypical total reduplication due to functional and formal aspects: syntactic dependency between base and copy
  A special case of non-prototypical reduplication is the so called contrastive focus reduplication. Here the copy is positioned left to the base and bears a focus accent, indicated by capital letters in this example from colloquial English: *"I'll make the tuna salad, and you make the SALAD-salad."* There is a dependency relation between base and copy, the latter acting as a modifier of the base and thus emphasising the essential meaning of the base on a more specialised range (Ghomeshi et al. 2004: 308).

1 A great exception is the Lithuanian case; Wälchli (2015: 504ff.) provides evidence that Lithuanian ideophones actually do show a modified meaning coming along with the reduplication.

The selected articles cover several linguistic areals from Southeast Asia to Africa and Europe and are dedicated to different types of non-prototypical reduplication. In alphabetical order, the contributors treat the following phenomena: Anvita Abbi gives a comprehensive overview of echo-word constructions and reduplicative expressives in Southeast Asia. Gregory D. S. Anderson presents an in-depth study on various reduplication phenomena – partial, total, and complex – in three languages from the Munda family. Rita Finkbeiner investigates a case of triplication in German which seems to be more systematic than simple repetition, discussing it within the framework of reduplication. Haritini Kallergi and Magdalene Konstantinidou provide a detailed insight into different kinds of echo-word constructions in Modern Greek, drawing parallels to diachronic data. Julia Nintemann deals with a formal aspect of reduplication proper in Bantu languages which poses a logical problem: the reduplication of monosyllabic bases. Fedor Rozhanskiy explores various (and typologically unexpected) patterns of Komi ideophones. Thomas Stolz contributes a thorough crosslinguistic investigation on reduplicative phenomena, testing both the prototype and the canonical approach and favouring the latter.

To all the workshop participants who for various reasons did not decide to include their paper in this volume I would like to express my deep gratitude – dear Ulrike Freywald, Rusudan Gersamia, David Gil, Zaal Kikvidze, Maia Lomia, Fabio Montermini, Hitomi Otsuka, Maja Robbers, Thomas Schwaiger and Giusi Todaro, your contributions were a substantial gain for the workshop and enriched the discussions. Sincere thanks belong to the student assistants who supported the meeting with their technical knowledge and cared for the snacks. My special thanks go to Dr. Cornelia Stroh who guided the publishing of this book along up to its finished state. All possibly remaining errors and inconsistencies are solely due to my lack of attentiveness.

Aina Urdze, Bremen, December 2017

References

Clark, Larry. 1998. *Turkmen reference grammar.* Wiesbaden: Harrassowitz.

Ghomeshi, Jila, Ray Jackendoff, Nicole Rosen & Kevin Russell. 2004. Contrastive focus reduplication in English (the salad-salad paper). *Natural Language & Linguistic Theory* 22. 307–357.

Otsuka, Hitomi. 2016. *Reduplikation in den Sprachen Japans (und benachbarter Regionen).* Bochum: Brockmeyer.

Rubino, Carl. 2013. Reduplication. In Matthew S. Dryer & Martin Haspelmath (eds.), *The world atlas of language structures online*. http://wals.info/chapter/27, accessed on 22.11.2017.
Southern, Mark R. V. 2005. *Contagious couplings: Transmission of expressives in Yiddish echo phrases*. Westport: Praeger.
Stolz, Thomas. 2008. Total reduplication vs. echo-word formation in language contact situations. In Peter Siemund & Noemi Kintana (eds.): *Language contact and contact languages. International Colloquium on Language Contact and Contact Languages. Hamburg, 06.–08.07.2006*, 107–132. Amsterdam: Benjamins.
Stolz, Thomas, Cornelia Stroh & Aina Urdze. 2011. *Total reduplication: The areal linguistics of a potential universal*. Berlin: Akademie Verlag.
Wälchli, Bernhard. 2015. Ištiktukai "eventives" – the Baltic precursors of ideophones and why they remain unknown in typology. In Peter Arkadiev, Axel Holvoet & Björn Wiemer (eds.), *Contemporary approaches to Baltic linguistics*, 491–521. Berlin: de Gruyter Mouton.

Anvita Abbi

Echo formations and expressives in South Asian languages

A probe into significant areal phenomena

Abstract: The paper discusses in detail echo formations and expressives found in South Asian languages. Their structural features and their associated semantics across a wide range of languages indicate that these two are significant South Asian areal phenomena (Emeneau 1956, 1969). The complex semantic structure of these constructions as well as a wide semantic and conceptual space they occupy – make them complex categories which had been ignored both by western and Asian linguists because they are a-prototypical grammatical feature not fitting the traditional grammar writing. Considering the effects of globalization, it is observed that while echo formations are stable structures which are indicators of social cohesion and bonding and thus find their inroads into ever-evolving languages, expressives are endangered structures and are dying very fast.

Keywords: semantically elaborate grammatical category, endangered structures, replacer syllable, reduplicator

1 Introduction

An echo formation (EF for short) can be defined as a partially repeated form of the base word where the reduplicator is a canonically copied form of the base with slight alternation. The alternation is brought about either by replacing the initial sound or initial syllable of the base word [reduplicant] by a replacer phoneme or a replacer syllable in the copied material. The entire construction of reduplicant and reduplicator constitutes an echo construction or echo formation. The copied element may be referred to as 'echo word'. Examples like Hindi *cay 'tea'* > *cay vay* 'tea etc./tea and related items'; *dal* 'daal' > *dal val* 'daal etc./daal and related items' are EFs and *vay* and *val* respectively are echo words. Let us consider the five prevalent ways of forming echo words across South Asian languages.

Anvita Abbi: Department of Linguistics, Simon Fraser University, Robert C. Brown Hall Building, 8888 University Drive, Burnaby, BC V5A 1S6 Canada. E-mail: anvitaabbi@gmail.com

https://doi.org/10.1515/9783110599329-001

2 The structure

1) Languages that allow the initial sound of the base word to be replaced by some specific sound which is unique to that language belong to the Indo Aryan, Indo-Iranian, and Munda languages. Thus, the Hindi word for 'song' is *gana* which can result in an echo construction as *gana vana* 'song and such activity', where the initial sound [g] of the base word has been replaced by [v] keeping intact the canonical form of the reduplicant. Kashmiri also uses *v-* as a replacer sound as in *pun vun* 'water etc.' In Kashmiri, if the word itself begins with a *v-*, the replacer sound assumes the shape of the bilabial plosive *p-*. In each language, there is some specific unique sound which plays the role of the replacer sound. Thus, this unique sound replacer is an identity marker of that language and there are several legends in each speech community which exposes interesting ways to find out the identity of a speaker by identifying the way s/he forms the echo word. Thus, Hindi has the replacer phoneme *v-* while Punjabi has *š-*, Andamanese Hindi has *m-* and Bangla has *ṭ-* and so on so forth (see Appendix 1).

It is to be noted that in Hindi if the base word begins with the specified replacer sound itself, i.e. [v], then the sound is dropped from the echo word, e.g. Hindi *vəkil əkil* 'lawyers etc.' (For details see Mohan 1999).

2) In Tamil, a Dravidian language, the word for 'tiger' is *puli* which becomes *puli gili* 'tiger etc.' after going through the EF process. The initial syllable *pu* is replaced by another syllable *gi* keeping intact the canonical shape of the base word. Thus, if the base word is /cv-x/ (where *cv* is the initial syllable of the word and the *-x* is the sequel), the EF of it would be /gi-x/. This replacer syllable /gi-/ is prefixed to all the echo words including those which have an initial open syllable. Thus, /v-x/ will result in a /v-x gi-x/ configuration, e.g., Tamil *uppu* 'salt' has the echo construction *uppu gippu* 'salt etc.' Telugu offers *a:ta* 'play' and *a:ta gi:ta* 'play and related activities'. The length of the initial vowel of the base word is copied in the replacer syllable. The rule applies even if the initial syllable is a consonant followed by a diphthong as the whole syllable is replaced by *gi*. For instance, Tamil *kəi:* 'hand' and *kəi: gi:* 'hand etc.' and Telugu *pəina* 'above' will be rendered as *pəi:na gi:na* 'above and related surface'. We can thus safely propose that the initial syllable of the base word in Dravidian languages is replaced by *gi* in echo word formation, copying the length of the vowel if any. Although Dravidian languages do not allow initial consonant clusters other than in those which are borrowed from other languages, the rule of echo word formation is maintained. Thus, the Sanskrit word *prema* 'love' becomes *prema gima* ‘love etc.’ in Telugu. The Dravidian echo word formation rule can be specified as follows:

$$\text{EF1} \qquad \# \, c_0^2 \qquad \underset{[\alpha\ \text{long}]}{\{v/sv\}} \longrightarrow g \qquad \underset{[\alpha\ \text{long}]}{i}$$

The South Indo-Aryan language such as Marathi uses the replacer syllable /bi/ as in Dravidian languages rather than one single sound to form an echo word. This can be ascribed to contact with Dravidian languages. Thus, Marathi uses *gel-a bil-a* ‘go-PST EW-PST’ ‘went (perhaps)’.

3) The third strategy is to have the reverse order of the reduplicant and reduplicator. It has been observed that some languages such as the Dravidian language Telugu, but not Tamil, offers not many but few examples of echo word preceding the base and not following it, unlike the prototypical pattern in South Asian languages. For example, the replacer syllable which is fixed in its form *-ṭṭa* is added to the first syllable of the base word and the whole echo word then precedes the base word. Consider:

(a) *paṭṭa pagalu* < *pagalu* ‘ broad day time’

(b) *baṭṭa bayalu* < *bayalu* ‘very open’

(c) *moṭṭa modalu* < *modalu* ‘right at the beginning’

Surprisingly, Indo-Aryan languages such as Hindi also allow a limited number of echo constructions where the echo word precedes the base, e.g. *bəgəl* ‘side’ > *əgəl bəgəl* ‘around’; *samne* ‘in front of’ > *amne samne* ‘in front of each other’; *pas* ‘near’ > *as pas* ‘near around, nearby’; *gine* ‘counted’ > *ine gine* ‘a few’; *pəṛos* ‘neighbor’ > *əṛos pəṛos* ‘neighborhood’ etc. As obvious in such constructions, the initial consonant is dropped while forming echo words.

4) There is a fourth strategy that is followed in forming echo words in South Asian languages, viz. vowel alternation of the initial syllable of the base word and copying the other elements of the word. This is seen more commonly in Western Indo-Aryan languages as well as in Indo-Iranian. Thus, the Punjabi word for ‘work’ is *kam* and the echo construction of the word is *kam-kum* ‘work etc.’ or in Bangani (an IA language spoken in the Himalayan region) the word for ‘meat’ is *šakun* and the echo construction is *šakun-šhukun* ‘non-vegetarian, meat related’. In Bhotia (also known as Jad) spoken in Dunda and Bagoshi in Uttarkashi (Uttarkhand) *ḍa ḍo* ‘rice etc.’ and *ŋalgo ŋulgo* ‘pillow etc.’ This strategy is not mutually exclusive to the first one. Many languages use both replacer sound scheme as well as vowel alternation. For instance, the example given above in Punjabi can be rendered alternatively as *kam-šam* ‘work etc.’ Hindi also offers vowel alternation as in *b*^{h}*ola-b*^{h}*ala* ‘innocent’, *d*^{h}*um-d*^{h}*am* ‘pomp and show’, *ʈ*^{h}*ik-ʈ*^{h}*ak* ‘arranged well’, *b*^{h}*iɽ-b*^{h}*aɽ* ‘crowd etc.’

5) The fifth strategy: Forming echo words by expressive morphology that I am going to discuss later in the paper. It will be safer to say that EFs are more an areal feature of languages belonging to the south of Himalayan region.

6) EFs are a **single lexical category** but not a single structural category. Each constituent of EF can take various inflectional and derivational affixes making them multiple structures. However, the whole construction stands for one specific meaning. This is obvious as the echo word is a copied element of the whole word with all its trappings. Thus, the Hindi word *gir-a-ya* ‘fall down-PST-3MSG.CAUS’ is rendered in EF as *gir-a-ya vir-a-ya* ‘made X fall down and such activities.PST-3MSG’. Another IA language, Marathi, represents a similar structure.

Marathi
(1) *kaɭ-i* *biɭ-i*
black-FSG EW-FSG
‘Black and dark shade’

7) The echo word has neither an individual occurrence nor any meaning of its own in the language. It acquires the status of a meaningful element only after it is attached to the base. In Abbi (1992) it is to be seen what kinds of elements are semantically encoded in the echo word or what is the range of the semantic field represented by echo constructions.

8) Functionally, all content words can be echo formed for all the meanings specified below. The rule applies to borrowed vocabulary too.

9) Languages differ as what is considered the base, with or without the inflectional/derivational affixes. While many IA languages allow affixes to be out of the copying rule and thus, affixes take the terminal position of the entire construction as in Bangla *mere-chi* ‘hit-PST’ > *mere ʈere-chi* ‘hit and related activity’, while others like Hindi copy the inflectional and derivational affixes of the base while constructing the echo word, e.g. *g*^{h}*ǝr-õ vǝr-õ mẽ* ‘in the house etc.’ or *gir-a vir-a* ‘fell and such activity’ or *sǝjav-ǝʈ vǝjav-ǝʈ* ‘decoration etc.’ Some Dravidian languages offer the option of either duplicating the bare form and adding affixes at the end of the echo word or duplicating the inflected word as in the case of Hindi. Thus, Kannada provides both options given below:

(d) *much gich-ide* ‘door EW-PST’ or *much-ide gich-ide* ‘ door-PST EW-PST’

Tamil, on the other hand does not allow the formation of echo words without the affixes. The rule is maintained even when there is a postposition and not a bound morpheme. Thus, in Tamil the base word is a word with all its bound morphemes including case marking in nominal constructions. Consider:

(e) **cennai kinnai-kku* 'Chennai EW-DAT'

(f) *Cennai-kku kinnai-kku* 'Chennai-DAT EW-DAT'
'To Chennai and some other city adjacent to it'

(g) **əppa kəppa-kiṭṭa* 'father EW-OBJ'

(h) *əppa-kiṭṭa kəppa-kiṭṭa keek-kaat-e*
father-at EW-at listen-NEG-IMP
'Do not listen to your father and other older people'

As said earlier, many IA languages such as Hindi and Marathi allow the fully inflected word as the base but when it comes to postposition the case marking is attached to the echo formation after the reduplication. Consider:

Hindi
(2) *gay vay ne cara k^ha-ya*
cow EW ERG fodder eat-3PST
'Cow etc./cattle has eaten their fodder'

and not

(3) **gay ne vay ne cara k^haya*

Although all grammatical categories can be used in EF, two echo formations consecutively cannot be used. Thus, the following is ungrammatical in Hindi:

(4) **pila vila dərwaza vərwaza*
yellow EW door EW

3 Semantics of echo constructions

It is not surprising how similar are languages of South Asia when it comes to representing the semantics of EFs. It is because of the shared structures and associated semantics that mark these structures as areal phenomena. The phonological echoing of the second part of the EF is intimately linked with the semantic echoing of the base word. This phenomenon is shared across languages.

One to one correspondence of the structure with its associated meaning gives us **system-iconicity** across languages.

Most importantly, EFs (i) encode one – or more – schematic meanings and (ii) may involve more than one conceptual-semantic domain. These can be regarded as **semantically "rich"** categories **or semantically elaborate grammatical categories** (Kuteva 2009; 2010) since they relate to more than one conceptual-semantic domain simultaneously. Accordingly, we will make a distinction between a universal conceptual-semantic space and a language-specific conceptual-semantic space.

Let us consider various shades of meaning underlying EFs. The semantic field of the EFs can be represented as a **cluster** of the following meaning components.

The foremost use of these constructions is to represent **generality and plurality** such as *pen-ven* in Hindi refers to related items like 'writing instruments'. All languages of South Asia below the Himalayan region –, i.e. excluding the Tibeto-Burman and Tai Kadai families, share this semantic construct by EF suggesting an areal nature of the construction.

Closely related to the concept of 'plurality' is the fact that the same EF may generate a **superordinate structure** enfolding many subordinate ones under it. For instance, in South Asian languages, there are no equivalents for English words like 'furniture' and 'stationary'. Names for any object belonging to these collective nouns can be echo formed to represent the superordinate structure, e.g., the same word *pen-ven* that we cited above in a different context means 'stationary items' and is not necessarily restricted to 'writing instruments'. This facility of creating a superordinate structure exists in all South Asian languages.

To **lessen the effect** of the base word especially when used in a verbal category, e.g. Hindi:

(5)	*uske*	*gane*	*vane*	*se*	*mujhe*	*koi*	*fark*	*nahĩ*	*paṭta*
	his	sing-INF	EW-INF	by	to me	any	matter	NEG	be

'His singing and such activity does not affect me'

(6)	*uska*	*likhna*	*vikhna*	*kisii*	*kaam*	*ka*	*nahĩ*
	his	writing-INF	EW-INF	any	worth	GEN	NEG

'His writing etc. [scribbling] is no good for me'

Contrary to above, EFs may be used to **increase the intensity** if echo construction is formed by vowel alternation. Hindi $d^{h}um$ $d^{h}am$ 'pomp and show', Punjabi *sida suda* 'absolutely straight' are some of the examples.

EFs **create sets and types** especially when modifiers are involved. Thus, Hindi *pila-vila* 'yellow types', *moṭa-voṭa* 'fat types', *b^hari-vari* 'heavy types', Punjabi *nila-šila* 'blue types' etc.

Ech formations are also used for **non-specific reference**, e.g. Hindi *dəs-vəs bəje* 'around ten o'clock', *kɛneda-vɛneda* 'Canada or some country in the west', *šila-vila ko bulao* 'call Sheela and any other girl', *pərsõ-vərsõ* 'day after tomorrow or later'. These kinds are used for hedging, when the speaker chooses to not to give full information. Ironically, the majority of the words under this category belong to proper nouns.

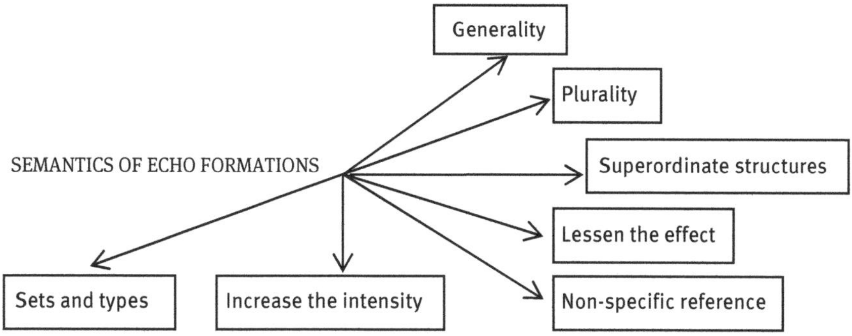

Figure 1: Semantic space of EFs.

EF is thus a **semantically elaborate grammatical category** for the reason that it encodes (1) generality, (2) plurality, (3) non-specificness, (4) accentuation, (5) attenuation, (6) creates superordinate structures and, (7) classify types.

The challenging task in front of us is to explain these grammatical polysemies. If we look closer we can, to a large extent, explain the polysemies by context sensitive rules. For instance, the matter of referring to 'superordinate structures' or to 'intensity' pertains to context sensitivity.

Can we identify a single coherent conceptual-semantic space represented by the EFs across various languages? The answer is in affirmative. If we define this space by a cluster of identifiable shared meaning components across languages, we can identify as well as organize the single coherent conceptual-semantic space.

4 Speech act phenomenon

EFs are heavily used in conversation for **pragmatic reasons** of making the addressee comfortable and relaxed. It is hence always a feature of informal spoken language. Because EFs are always part of informal conversation they are avoided when the dyads are in asymmetrical relation. Thus, a student will avoid using these while talking to his teachers and elders, while the teacher and elders may use these constructions without any hesitance.

Two important conclusions can be drawn here: first, the use of echo word formation presupposes sharing of a cognitive environment and a conceptual space by the speaker and the hearer. Unless the cognitive environment and the conceptual space is shared, the range of meanings and associated semantics of each of the EFs will be misinterpreted leading to communicative chaos. Secondly, the use of EFs marks the social solidarity and mitigates differences. It is a big equalizer among the disparate groups. One never uses these constructions with strangers no matter how good the competency is of the hearer in the language.

To put it differently, EFs are complex and semantically elaborate grammatical categories. The utterances marked by EF are contextually determined for their semantics and construct reality of 'equality' or 'fraternity' and 'comradeship' among the interlocutors. These are significant linguistic structures which are inherent components of language acquisition to function in the South Asian society. Words not only refer to objects and actions, they can serve as markers of social cohesion and bonding. It is in this area that EFs operate.

Seen from the point of view of the areal aspect, the shared semantic field/s represented by the same structure, viz. echo word formation, indicates shared underlying cognitive ability, sharing of the single coherent conceptual-semantic space and common interpretation across the speakers of diverse languages of South Asia.

5 Expressives

Now I move on to a phenomenon not new but still challenging in linguistic theory and literature as it has not found its due place in any grammatical theory. This is because of our obsession with restricting to prototypical categorizationalists' approach of Aristotelian categories. We must admit that categories may encode one or more schematic meanings across different conceptual-semantic domains (Kuteva 2009; 2010). Categories can be structurally definable and semantically complex. Categories can occupy a major grammatical structure of a language without being a typical noun, adverb, adjective or verb. The structure I am referring to are expressives. Unfortunately, these are considered perfunctorily in any grammar if we are lucky enough to find their mention. Most often than not, expressives are unsung heroes of any grammar of South Asian languages. The form-meaning pairing that we review here presents a challenge exactly because of its complex semantics.

5.1 The term

The term 'expressives' as used in this paper is inclusive of ideophones, onomatopoeics, mimics, imitatives, and sound symbolism (for details on each of these to avoid the confusion created by the overlapping definitions, see Abbi 1992). Although many writers today use the term 'expressive', there has been something of a naming frenzy in the past. In earlier works, especially on African and South Asian languages, expressives have also been given labels such as 'interjections', 'descriptive adverbs', 'picture words', 'adverbials', 'intensives', 'emphatics', 'impressifs', and so on. Again, different scholars give different definitions of expressives.

According to Childs (1989: 1), the term 'expressives' seems to have been first coined by Durand (1961) in his analyses of Vietnamese. The term was later adopted and defined by Diffloth (1972, 1976) and Emeneau (1978). The term 'ideophone' is widely used, however, for the African phenomenon, as in Doke (1935) for Bantu. He seems to have first suggested the term, he defined or at least described an ideophone as a vivid representation of an idea of sound, a word, often onomatopoeic, which describes a predicate, qualificative or adverb in respect to manner, color, smell, action, state or intensity. This seems to suggest that ideophones are a grammatical class of words, a type of adverbial, but no formal criterion is given for distinguishing them from other adverbs. In the Bantu languages with which Doke was concerned, the invariable or indeclinable nature of ideophones had often been noted, in contrast with those adverbials formed with the locative class prefixes.

It has also been noted that expressives, for some languages, are frequently phonologically anomalous. They may contain phonemes not found in other types of words, or unique sequences of phonemes, and they may be errant in respect to the rules of tone that apply to them. Diffloth holds a similar opinion

(1972).[1] Diffloth (1976: 249), for Semai, considers expressives as a “third basic word class” of the same order of magnitude as nouns and verbs. Describing the morphological, syntactic and semantic properties of Semai expressives, he proves that they are not even subject to the condition of ”lexical discreteness” and are indeed ”a totally different kind of linguistic animal”.

Diffloth (1972: 442) very strongly advocates the iconicity point of view of expressives, “iconicity is the very *raison d'être* of the whole word class called Expressives”, as well as suggests that the linguistic theory has to be overhauled completely to incorporate this phenomenon, “it will be necessary to create an **aesthetic component** (our emphasis) of grammar, distinct from, but incorporated into, the logicophonological component which has been the sole preoccupation of generativists”.

Emeneau (1978) in his study of Kota onomatopoeics, opines that expressives can be identified by having syntax and morphology different from that of the major classes of the language. Abbi (1987) rejects such a position. For her, expressives do not necessarily have distinct morpho-syntactic properties than the other lexical items in the language. For example, verbal onomatopoeics and imitatives in Hindi and Tamil operate as normal verbs and take normal affixations allowed in the languages. Emeneau (1978), however, rightly comments that “...perhaps it would be more just to say that expressives denote varied types of sensation, the impingement of the material world, outside or within the person, upon the senses-not merely the five conventionally identified senses, but also the feelings both internal and external”.

5.2 The structure of expressives

Abbi (1987, 1990, 1992) discusses expressives from the point of view of reduplication, the linguistic structuration, which is more common and widespread in all the South Asian languages than the non-reduplicated ones. Whether the expressive form is reduplicated or non-reduplicated, each form is a unit lexeme and a single structural category. Almost all (nearly 99 %) reduplicated expressives are formed by iterating a meaningless syllable. The resultant structure thus acquires a meaning, constitutes a single morpheme as well as a single lexeme in that language. Thus, Sora (Austroasiatic) *mel mel* ‘to inspect’, *di di* ‘to count’, Khasi *raʔ raʔ* ‘flowers’ are words derived by expressive morphology. An expressive derives its status of a word/lexeme only after it is duplicated, as the

1 The author disagrees with this view as it is discussed later in the paper.

non-reduplicated syllable does not exist as a word. However, although small in number, Naga languages (Tibeto-Burman) do have non-reduplicated expressives. Abbi (1987) considers all expressives as instances of morphological reduplication as opposed to the lexical reduplication where the units before iteration are meaningful words of the language concerned. The question of expressives being iconic cannot be established without some doubt. Had these been totally iconic, languages of a *sprachbund* would share their phonetic shapes of expressives. Nonetheless, some sound symbolism is involved but no one can be sure about their total iconic nature. For instance for 'rain pattering sound' Hindi has *ʈəp ʈəp* while Mizo, the Tibeto-Burman language has *klɛk klɛk*; for sense of sight Hindi has *cam cam,* Nepali uses *bʰər bʰər* and Mizo has *sɛp sɛp*. Consider Appendix 2 on expressives to see the variety of forms to denote universal conceptual semantic space.

As mentioned earlier, expressives behave and function like regular words and thus form a part of the lexicons of Indian languages. Unlike many other languages of the world, expressives in Indian languages can form the predicates. The morphological form of the expressive word varies from language family to language family. It can be suffixed by a conjunctive participle in Indo-Aryan and Dravidian, while in Tibeto-Burman it can be prefixed by a particle indicating 'manner'. It constitutes as a bare adverbial category and thus does not take any suffix/prefix in Khasi, the Mon Khmer language of the Austro-Asiatic family.[2] Before we proceed, let us inform our readers that it is dauntingly difficult to translate expressives in English and several explanations at best, prove to be at times, marginally close to the real meaning. See examples given below.

Expressives used in various morphological paradigms in Hindi:

Root	*pʰəʈ*	semantically vacuous
Stem	*pʰəʈ pʰəʈ*	'flutter'
Infinitive	*pʰəʈ pʰəʈ -ana*	

	PAST	PRS.IMP	FUT
3MSG	*pʰəʈ pʰəʈ -ya*	*pʰəʈ p pʰəʈ -ta*	*pʰəʈ pʰəʈ -yega*
3FSG	*pʰəʈ pʰəʈ -yi*	pʰəʈ *pʰəʈ -ti*	*pʰəʈ pʰəʈ -yegi*
3MPL	*pʰəʈ pʰəʈ -ye*	*pʰəʈ pʰəʈ -te*	*pʰəʈ pʰəʈ --yẽge*
3FPL	*pʰəʈ pʰəʈ -yĩ*	*pʰəʈ pʰəʈ -tĩ*	*pʰəʈ pʰəʈ -yẽgi*

Derivative Noun	*pʰəʈ pʰəʈ-ahəʈ*	'fluttering/flutter'
Derivative Modifiers	*pʰəʈ pʰəʈ -ay-a/i-hu-a/i*	'having being fluttered 3MSG/3FSG'

2 In case of Khasi, these constructions are not verbal adverbs but adverbs per se.

5.2.1 Grammaticality

Expressives can function as typical nouns or verbs of the language concerned. They are employed in regular grammatical paradigms of the language and thus form an integral part of the lexicon. In other words, they do not necessarily have morphosyntactic characteristics distinct from the rest of the lexical items in these languages. For instance, many of the imitative expressives (acoustic noises) in Hindi operate as an ordinary verbal category taking the usual affixes of Hindi.

It has been observed by Abbi (1994) that some expressives have their origin in words which are cognates across the same language family. Thus, Hindi *təm təm-ana* 'to redden with anger' is derived from the word *tamra* 'copper colored, copper' which finds its cognates in Pali *tamba* 'red', 'copper'; Prakrit *tamba* 'red (adjective/noun)', Dameli *trāmba* 'red' and Sinhalese *tambha* 'reddish'. It is to be established historically whether expressives are derived from verbs or verbs are derived from expressives after de-expressivizing them. Okombe-Lukumbu Tassa (in Voeltz & Kilian-Hatz 2001) cites both the processes for Tetela (Bantu).

5.2.2 Semantics of expressive morphology

As said earlier, expressives represent a complex semantic category in all languages of South Asia. Their complexities have been the sole driving force for not being recognized as the inherent part of any grammar. We shall now explain these.

5.2.2.1 Complex semantic category

Expressives represent
- Five senses of perception (*panchendriya*)
- States of mind
- Manner of an action
- Kinship terminology – (language universal)
- Various states of confusion

We shall discuss each of them in the following pages.
1. Five senses of perception (*panchendriya*)
   We must inform that the most significant aspect of Indian expressives is that they, without fail, indicate five senses of perception, viz. of smell, sight, touch, hearing and taste. Thus, *gam gam* 'aroma' in Maithili, *cam cam* 'glit-

tering' in Hindi, *las las* 'sticky' in Punjabi, *khe khe* 'laughter' in Meitei, *ṭok ṭok* 'laughing sound' in Kurukh and *kur kur-a* 'crunchy' in Hindi are typical examples of expressives indicating distinct five senses of perception. See Appendix 2 for a wide range of expressives in Indian languages.

2. As 'manner' of an action
   Another semantic area most widely covered by expressives is that of the 'manner' of an action/event stated. Every society and its members perceive the world distinctively. The perception of the manner in which any activity can be undertaken is culture-specific. That is, each culture has its own views and models the way any activity is undertaken. Society and culture-specific 'manner' in Indian languages are expressed largely by expressive morphology than by prosaic words of manner adverbs. The phenomenon is in greater abundance in the indigenous languages of India than in the modem languages and, among the indigenous ones, more prevalent in the languages of the Northeast, both in Tibeto-Burman, Tai-Kadai and Khasi (Austroasiatic). We shall dwell in detail on this area to highlight this important phenomenon of micro area. Many in literature had classified ideophones as a subclass of adverbs (Schaefer 2001).

5.2.3 Himalayan region of South Asia

Although expressives is a common phenomenon in South Asian languages, it is abundant in languages spoken in the Himalayan region be it of Tibeto-Burman language family, or Tai-Kadai, or Mon Khmer branch of Khasi, or Western Himalayan languages of Indo-Aryan. We will thus draw our examples primarily from these languages. We would like to take into consideration the languages of the Northeast, primarily Naga languages of the Tibeto-Burman family, Meitei of the Kuki-Chin group, Tai-Khamti from Tai-Kadai and Khasi of the Mon-Khmer group of the Austro-Asiatic language family. These examples represent the range of perceptive powers of the speech communities of Naga, Kuki-Chin, Khasi and Meitei. We identified 59 expressives, all indicating the manner of walking in Khasi and an equal number in Tangkhul Naga (Abbi & Victor 1997). Action verbs such as 'crying', 'walking', 'running', 'laughing' etc. are coded with a high number of expressives.

5.2.3.1 Tangkhul Naga and Meithei (Tibeto-Burman)

Tangkhul Naga belongs to the Kuki-chin branch of the Tibeto-Burman language family and is spoken by 1,500,000 people who are spread over the whole Ukhrul District of Manipur – a hilly terrain spreading over 1823 sq. miles. A large number of the speakers of this language is scattered outside Ukhrul District.

The language is known by their ethnic name 'Tangkhul Naga'. It consists of many dialects – each village has its own dialect named after the village. The intelligibility among the village dialects varies according to the distance between them. The farther the village, the more is the unintelligibility.

Out of 342 expressives known so far in Tangkhul Naga, 278 begin with consonants, 31 with vowels and 33 with semi-vowels. That is, about 80 % of the expressives have initial consonants. Further classification shows that out of 278 expressives with initial consonant, 48 are non-reduplicated and the rest are reduplicated either partially or completely (Victor 1992). Phonologically, partial reduplication undergoes a process in which a vowel or a consonant/semi-vowel or both are changed, deleted or added in the reduplicated part.

As in other South Asian languages Tangkhul Naga expressives can occupy both the verbal and the adverbial slots, meaning thereby that expressive either forms the predicate or occurs in adjunct position as a verbal modifier. The prefix *ta-* is used as an adverbial particle to the expressive. It is noteworthy that this prefix cannot be attached to any other word class. This is discussed below under 'syntactic characteristics'.

We shall cite some examples from the reduplicated expressives as they are more commonly found than the non-reduplicated ones. Reduplicated expressives are either partially or fully reduplicated, the latter ones outnumber the former. Interestingly, in partial reduplicated form, the vowel is altered along with the tone. Hence the reduplicator and the reduplicant have different tone markings. Consider one such example.

o ——→ ***a***

(i) *ho?ó:ho ho?à:ho* 'expression used in quieting babies or putting them to sleep'

More often than not, reduplicated expressives have tone, marked only once either on the repeated part or on the base. Thus:

(j) *ŋə̀ŋ ŋəŋ* 'burning of a huge fire producing a lot of noise, high flames and consuming fuel wood very fast'

However, cases of both reduplicator and reduplicant being marked by tones are not unusual. Consider:

(k) *tít tít* 'to be very tight like a gunny bag due to over-stuffing, or a garment to be very tight when put on'

(l) *rór rór* 'do something one after another repeating the same action'
[Victor 1992: 48]

Various kinds of syllables take part in forming expressives, monosyllabic, bisyllabic and trisyllabic base can be reduplicated. Similarly, initial sound can be a vowel, a consonant or a semivowel. Tones can be variable or fixed. All permutations are possible as the language is very rich in its stock of expressives.

Another interesting feature of this language is that expressives like any word can be compounded, i.e. two expressives can form a **compound**. Compounds can be either compositional, i.e., retaining the individual meaning of each expressive or opaque referring to some third entity. Structurally, various ways can be employed to derive an expressive compound, viz. two bisyllabic reduplicated expressives can be compounded, or one of the syllables can be dropped, or tones can change resulting in Tone Sandhi as shown in the following table. Tangkhul Naga is the only language that allows three expressive word compounds. Consult Table 1 given below.

Compounding results in tonal change. Consider:

(m) *yáŋ + yiŋ̀ = yaŋ̀ -yiŋ* 'the state of having a relieving sigh, fragile, touchy moving (emotionally), orchids'

(n) *yáŋ* 'sudden short spanned emotional feeling as happiness or sadness'

(o) *yiŋ̀* 'emotional disturbance caused by the sight of extremely beautiful or very bad colors or looks'

The high-toned *yáŋ* becomes low-toned and low-toned *yiŋ̀* becomes mid-toned when compounded.

Table 1: Expressive compounds in Tangkhul Naga.

No.	Expressive 1	Expressive 2	Compound	Remarks
1	*yáŋ* 'sudden short spanned emotional feeling as happiness or sadness'	*yiŋ̀* 'emotional disturbance caused by the sight of extremely beautiful or very bad colors or looks'	*yaŋ̀-yiŋ* 'the state of having a relieving sigh, fragile, touchy moving (emotionally), orchids'	Derived out of non-reduplicated expressive. The high-toned *yaŋ* becomes low-toned and low-toned *yiŋ* becomes mid-toned when compounded

No.	Expressive 1	Expressive 2	Compound	Remarks
2	*pʰút pʰut* 'do something with special need or urgency'	*nám nam* 'go in a rush manner and straight not caring for hindrances'	*pʰút pʰut nám nam* 'do, go, act quickly without pause or questioning as in case of emergency'	Compositional meaning is retained and original tones are also retained
3	*zí zí* 'moving here and there restlessly, dropping and peeping'	*par par* 'touching everything at reach in a hurried manner as if looking for something urgently needed'	*zí zí par par* 'having a special work needing close attention and anxiety for its completion, especially at the tilling and transplantation season'	Complete reduplication with original tones being maintained. Semantic shift as meaning is no longer compositional
4	*yúp yup* 'becoming darker and darker very fast as with burning out lamps'	*yəp yəp* 'instinct sight of momentary appearance and disappearance of objects'	*sə́p səp* 'doing or saying anything, moderately, or neither less nor more'	*yupyəpsəp* 'time span of one hour just after the sunset' Remarks: three expressive compound with truncated second syllable and loss of tones

For details on their formation and semantics one should consult Victor (1992).

5.2.3.2 Khasi (Austroasiatic, Mon Khmer)

During our fieldwork conducted in 1997 we came across Khasi expressives in large number where we identified 66 different expressives that collocate with the verb *yaid* 'walk'; 57 which collocate with the verb 'cry', 20 expressives collocating with the verb *ba:m* 'eat'; 28 expressives with the verb *khin* 'tremble'; 38 with the verb *krin* 'speak', 'say' and 18 expressives collocate with the verb *mareh* 'to run'. The largest number of expressives begins with *k-* or *kh-* in Khasi. The enormity of manner expressives collocating with the verb 'to walk' in the Northeast could certainly license the area as a 'Walking area' (Abbi & Victor 1997: 427). This is illustrated in Appendix 3.

5.2.3.3 Tai-Khamti (Tai-Kadai)

A very productive use of expressive morphology is made for accentuation of any color or for accentuation of any attribute. Consider the following examples (Sharma 2014):[3]

(p) *syen*^{4} *–sok*^{2}*-sok*^{2} 'very beautiful'
suŋ^{1} *–wen*^{2}*-wen*^{2} 'very tall'
suŋ^{1} *ngau*^{2}*-ngau*^{2} 'very tall'
təm^{1} *–me*^{6}*-me*^{6} 'very short'
cəm^{3} *thɔk*^{6}*-thɔk*^{6} 'very close'
yəm^{5}*sɔk*^{2}*sɔk*^{2} 'wet completely'

5.2.3.4 Nepali

Interestingly, South Asian expressives are not restricted to the perceptual words and manner of actions alone. There are many in various languages which lay bare many feelings, situations, attributes, disorder, disturbance, confusion, untidiness and, as Emeneau (1978) says, "our internal and external feelings".

Nepali [Abbi 1992: 18]
(q) *k*^{h}*əl bəl* 'hurly burly', 'commotion'
k^{h}*əla bɛla* 'riot uproar'
c^{h}*ul c*^{h}*ul* 'unsteady'
k^{h}*oṭəl k*^{h}*əṭəl* 'topsy turvy'
əṭ pəṭ 'disorderliness'

As far as the kinship terminology is concerned, not all but the majority of the languages of the world derive their kinship lexicon by expressive morphology, primarily by duplicating the initial syllable, e.g. *mama* 'mother', *papa* 'father' in English; *dada* 'grandfather', *didi* 'elder sister' in Hindi, also with vowel alternation such as *dadi* 'grandmother' (father's mother) and *nani* 'grandmother (mother's mother)'.

As we noted above in Section 5.2 expressives at the morphological level can work as any other word in the language. The following section illustrates the syntactic characteristic of these constructions.

What strikes one is the use of the similar linguistic material for shared semantic constructs by these languages. The structural similarities are the significant linguistic truths of the area (for detail see Abbi 1993).

3 Numbers indicate the tone levels.

Thus, the expressive word constitutes a single coherent conceptual-semantic space across the languages of the Himalayan region. This semantic space includes definable but semantically complex categories. However, expressives in general – considering the South Asia as a single linguistic area – are categories which encode one or more schematic meanings across different but inter-related conceptual-semantic domains.

5.2.3.5 Syntactic characteristics

Expressive verbs have the following characteristics at the syntactic level in most of the South Asian languages:

- They have the potentiality of being used as a finite verb, e.g. Hindi

  (7) *use* *dekhte* *hi* *mɛ̃n* *thərthər-aya*
  3SG-ACC see-PRS.IMP.OBL EMPH 1SG EXPR-3MSG
  'I shivered seeing him.'

- They have the potentiality of being used as a conjunct verb/converb, e.g. Hindi:

  (r) d^{h}ək *d^{h}ək kər (do)/ hona (be)* 'throbbing'.

- They have the potentiality of being used as a complex predicate in conjunctive participle form, such as in *d^{h}ək d^{h}ək kər ke (*EXPR + do + CP) 'having throbbed' or 'throbbingly', which serves as a manner adverb.
- In Tangkhul Naga (and other Tibeto-Burman languages) they can be prefixed by *ta-* to mark the adverbial or adjectival nature to the construction, e.g. *ta-yok yok* 'go wearily'; *ta-yok yok hay-rə* 'he is weary'. An important feature about the expressives is that like any other modifier category, they have their own strict collocational restrictions – each type is rule governed to appear with a select few verbs and nouns.

Unlike ideophones in African languages (Newman 1968: 107–118) Tangkhul Naga expressives can occur in all sentence types – declarative, imperative, potential, obligatory, conditional, permissive, aphoristic, vocative and topicalized/focused (Victor 1992: 69). Consider a few examples:

tham tham = loudly, with a lasting impression, do something with force without fear and hesitation

(8) [declarative]
a ***tham tham*** *mətuy-tə* *ləy*
3MSG EXPR speak-VPT PRS
'He is speaking loudly without any hesitation'

(9) [imperative]
tham tham *khə-mətuy* *təm-lu*
EXPR INF-speak learn-IMP
'[You] learn to speak loudly and impressingly without fear and pause'

(10) [potential]
i *zət-khəreotə* *a* ***tham tham*** *mətuy-phokhauləpay*
1SG go-as soon as 3SG EXPR speak-start may
'As soon as I leave he may start speaking loudly'

(11) [optative]
ithum-wuy *məśun* *vaŋ* ***tham tham*** *mətuy-sə*
we-GEN right for EXPR speak-OPT
'Let us speak without fear for our rights'

(12) [conditional]
tham tham *mətuy-əkhə* *mipiŋ-nə* *nə-li* *so-rə*
EXPR speak-COND people-NOM you-ACC praise-FUT
'If you speak impressively people will praise you'

Tangkhul Naga is a language, which does not have a separate word class of 'adverbs'. It has four different adverbial affixes that are affixed to adjectives to produce 'adverbial meaning'. The expressive is the only way the language generates adverbs.

5.2.3.6 Some examples of expressives used as finite verb, manner adverb and with conjunctive participle in other languages

These are extremely common in Bangla. Consider the following examples from Indo-Aryan and Dravidian:

Bangla
(13) *śe* ***huʈmuʈ*** *korch-e*
3SG EXPR hurried do PRS-3SG
's/he is hurrying up'

(14) *mey-ʈi* ***dum dam kɔre*** *kɔtha* *bɔle*
girl-CL EXPR-CP talk speak
'The girl speaks without thinking'

Angika
(15) *matha* ***hən hən-awe-che/*** ***hən hən*** *kər-eche*
head EXPR-3SG-PRS/ EXPR DO3SG-PRS
'Head is aching with heavy and pulsating sense'

(16) Tamil

(s)	*taṭataṭa-ttal*	'to walk unsteadily with age, to be loose'
	taḷutaḷu-ttal	'to falter or stammer from ecstatic joy, love or other emotion'
	tiṭutiṭu-ttal	'beating of the heart with fear, speedily'
	naḷunaḷu-ttal	'speak evasively'
	neruneru-ttal	'feel sudden pain as in the stomach'
	noṭunoṭu-ttal	'to be fidgety, restless, to be rude'
	koḷukoḷu-ttal	'to become loose, deranged

All these characteristics draw our attention to the complexity of the grammatical category named expressive, surely warranting a definite place in the grammar. Expressives carve out an independent category status in any South Asian language because of its morphological, syntactic and semantic behavior.

5.2.3.7 The social aspect

As in the case of EFs, expressives assume a significant part in social cohesion and solidarity. As Childs (2001: 66) puts it "Ideophones are quintessentially the mark of local identity but an identity of continuity with an ideophone past". South Asian communities regard the knowledge of expressives as a marker of intelligence and identity both. There are innumerable numbers of fables and folk tales as how a stranger in the village was identified as soon as he started speaking with wrong set of expressives, or divulged his identity while muttering in dream of unknown expressives. Bhaskararao (1977: 31) quotes an interesting fable from Telugu, a Dravidian language where the king tests the intelligence of his minister by giving only a series of expressives *puṭukku jarajara ḍubbukku me:* 'the goat bleated after the dry bottle gourd fell on it'.

6 Conclusion

There is no doubt that because of unusual nature of these constructions – both EFs and expressives – have escaped the kind of attention they deserved in the grammar. The Eurocentric and Westernized linguistic theories avoided establishing a proper place for these constructions in grammatical theories. This could be because of the complex semantic structure of these constructions as well as a wide semantic and conceptual space they occupy – the phenomena not very trivial to handle in a systematic way; or, because these did not fit into the Aristotelian categories of grammar. The Asian linguists ignored them from describing and giving them a due place in descriptive grammars because of the social nature of these constructions. These are, more often than not, spoken phenomena and are used in informal settings. These constructions never find their place in written literature other than in the genre of short stories and poetry. Perhaps these were the reasons for their omission from the grammatical description of any language. What Watson (2001: 401) said of African languages is so very true for South Asian languages, viz. these have been the victims of "textual genocide".

What worries me most is the fact that while EFs are stable structures which find their inroads into ever-evolving languages, expressives are endangered structures and dying very fast. The examples that I have quoted here from walking expressives (Appendix 3) are no longer found in the speech of the young generation of Khasi. These are losing at an alarming speed. Considering the fact these are heavily loaded semantically and culturally, loss of these structures will have a serious impingement on the cognitive abilities of the community.

Some of the issues and questions that we can address to in future are:

- Can these structures be considered the result of a co-evolutionary approach where culture plays the centre stage?
- Do these structures confirm the theory of relativity, i.e., do these structures determine the perception of modes of an action?
- Or conversely, are these structures the reflection of the culture-specific society that has heightened perceptive capabilities?
- How to assign an appropriate place of this complex category in grammar as it occupies wide semantic and conceptual space?

We hope to have answers to some of the questions when we meet next.

Abbreviations

1, 2, 3	first, second, third person
ACC	accusative
CAUS	causative
CL	classifier
COND	conditional
CP	conjunctive participle
DAT	dative
EF	echo formation
EMPH	emphatic

ERG	ergative
EW	echo word
EXPR	expressive
FSG	feminine singular
FUT	future
GEN	genitive
IMP	imperative
INF	infinitive
MSG	masculine singular
NEG	negative
NOM	nominative
OBJ	object
OBL	oblique
OPT	optative
PRS	present
PST	past
SG	singular
VPT	verbal particle

References

Abbi, Anvita. 1987. Semantic correlates of the Indian linguistic area: A study in reduplicative structures. In Elena L. Bashir, Madhav K. Deshpande & Peter E. Hook (eds.), *Select papers from SALA-7: South Asian Languages Analysis Roundtable Conference held in Ann Arbor, Michigan, May 17–19, 1985,* 1–12. Bloomington, Ind.: Indiana University Linguistics Club.

Abbi, Anvita. 1990. Reduplication in Tibeto-Burman languages of South Asia. *South East Asian Studies* 28(2). 171–181.

Abbi, Anvita. 1992. *Reduplication in South Asian languages: An areal, typological and historical study*. New Delhi: Allied Publishers.

Abbi, Anvita. 1994. *Semantic universals in Indian languages.* Simla: All India Institute of Advanced Study.

Abbi, Anvita & Ahum Victor. 1997. Expressive morphology and manner adverbs in Khasi, Tangkhul Naga and Kuki-Chin languages. In Anvita Abbi (ed.), *Languages of tribal and indigenous people of India: The ethnic space*, 415–431. New Delhi: Moti Lal Benarsi Das.

Bhaskararao, Peri. 1977. *Reduplication and onomatopoeia in Telugu.* Poona: Deccan College.

Childs, Tucker G. 1989. Where do ideophones come from? *Studies in the Linguistic Sciences* 19. 57–78.

Childs, Tucker G. 2001. Research on ideophones, whither hence? The need for a social theory of ideophones. In F. K. Erhard Voeltz & Christa Kilian-Hatz (eds.), *Ideophones. Typological studies in language*, 63–73. Amsterdam/Philadelphia: John Benjamins.

Diffloth, Gérard. 1972. Notes on expressive meaning. *Chicago Linguistics Society* 8. 440–447

Diffloth, Gérard. 1976. Expressives in Semai. In Philip N. Jenner, Lawrence C. Thompson & Stanley Starosta (eds.), *Austroasiatic studies.* Part 1, 249–264. Honolulu: University of Hawai'i Press.

Doke, Clement M. 1935. *Bantu linguistic terminology*. London: Longmans, Green and Co.
Durand, Maurice. 1961. Les impressifs en Vietnamien. Étude préliminaire. *Bulletin de la Société des Études Indochinoises* 36(1). 7–50.
Emeneau, Murray B. 1956. India as a linguistic area. *Language* 32. 1–17.
Emeneau, Murray B. 1969. Onomatopoeics in the Indian linguistic area. *Language* 45. 274–299.
Emeneau, Murray B. 1978. Some notes on Dravidian intensives. In George Cardona & Norman H. Zide (eds.), *Festschrift for Henry Hoenigswald: On the occasion of his 70th birthday*, 109–113. Tübingen: Gunter Narr. [Repr. In Emeneau 1994: 323–327].
Kuteva, Tania. 2009. Grammatical categories and linguistic theory: Elaborateness in grammar. In Peter K. Austin, Oliver Bond, Monik Charette, David Nathan & Peter Sells (eds.), *Proceedings of Conference on Language Documentation and Linguistic Theory 2*, 13–28. London: SOAS.
Kuteva, Tania. 2010. Grammaticalization and the evolution of grammar: On one particular kind of grammatical categories. In Young-Se Kang, Jong-Yuri Yoon, Jongseon Hong, Jiun-Shiung Wu, Seongha Rhee, Kyoung-Ae Kim, Dong-Ho Choi, Kee-Ho Kim & Hye-Kyung Kang (eds.), *Lectures on universal grammar and individual languages (SICOL-2010)*, 144–156. Seoul, Korea: Hankookmunhwasa.
Mohan, Shailendra. 1999. Morpho-semantic study of echo words in Hindi. Case of speech act phenomenon. M.Phil. Dissertation. Jawaharlal Nehru University. New Delhi. Unpublished.
Newman, Paul. 1968. Ideophones from a syntactic point of view. *Journal of West African Languages* 5. 107–118.
Schaefer, Ronald. 2001. Ideophonic adverbs and manner gaps in Emai. In F. K. Erhard Voeltz & Christa Kilian-Hatz (eds.), *Ideophones. Typological studies in language*, 339–354. Amsterdam/Philadelphia: John Benjamins.
Sharma, Bishakha Das. 2014. A descriptive grammar of Tai-Khamti. Ph.D. Dissertation, JNU, New Delhi.
Tassa, Okombe-Lukumbu. 2001. La formation des radicaux déidéophoniques et des idéophones déverbatifs en tɛtɛla (dialecte ewango). In F. K. Erhard Voeltz & Christa Kilian-Hatz (eds.), *Ideophones. Typological studies in language*, 375–384. Amsterdam/Philadelphia: John Benjamins.
Victor, Ahum. 1992. A grammatical study of expressives and echo formation in Tangkhul Naga. M.Phil. dissertation. Jawaharlal Nehru University. New Delhi. Unpublished.
Voeltz, F. K. Erhard & Christa Kilian-Hatz. 2001 (eds.). *Ideophones. Typological studies in language*. Amsterdam/Philadelphia: John Benjamins.
Watson, Richard L. 2001. A comparison of some Southeast Asian ideophones with some African ideophones. In F. K. Erhard Voeltz & Christa Kilian-Hatz (eds.), *Ideophones. Typological studies in language*, 385–405. Amsterdam/Philadelphia: John Benjamins.

Appendix 1

Table 2: Nominals.

No.	Language	Word	Echo Word Formation
1.	Tamil	*puli* 'tiger'	*puli-**gili*** 'tiger, etc'.
2.	Telugu	*puvu* 'flower'	*puvu-**givu*** 'flower etc'.
3.	Hindi	*pʰul* 'flower'	*pʰul-**vul*** 'flower etc'.
4.	Marathi	*kʰoli* 'room'	*kʰoli-**bili**/* 'room etc'.
5.	Kannada	*pennu* 'pen'	*pennu-**ginnu*** 'pen etc'.
6.	Bangla	*baɽi* 'house'	*baɽi-**taɽi*** 'house etc'.

Table 3: Adjectives.

No.	Language	Word	Echo Word Formation
1.	Tamil	*pəcca* 'green'	*pəcca-**gicca*** 'green and the life'
2.	Telugu	*cinna* 'small'	*cinna-**ginna*** 'small etc'.
3.	Hindi	*moʈa* 'fat'	*moʈa-**voʈa*** 'fat etc'.
4.	Marathi	*kaLa* 'black'	*kaLa-**biLa*** 'black or of dark complexion'
5.	Kannada	*doḍḍa* 'large'	*doḍḍa-**giḍḍa*** 'large and like'
6.	Bangla	*kalo* 'black'	*kalo-**ʈalo*** 'black and like'

Table 4: Verbals.

No.	Language	Word	Echo Word Formation
1.	Tamil	*vəndu* 'come'	*vəndu-**gindu*** 'come etc'.
2.	Telugu	*vintəm* 'hear'	*vintəm-**gintəm*** 'hear etc'.
3.	Hindi	*jana* 'go'	*jana-**vana*** 'go etc'.
4.	Marathi	*gela* 'go'	*gela-**biḷa*** 'went etc'.
5.	Kannada	*ooda* 'run'	*ooda-**giida*** 'run etc'.
6.	Bangla	*mere* 'beat'	*mere-**ʈere*** 'beat etc'.

Table 5: Adverbials.

No.	Language	Word	Echo Word Formation
1.	Tamil	*metuvaa* 'slowly'	*metuvaa-**kituvaa*** 'slowly etc'.
2.	Telugu	*tondərga* 'fast'	*tondərga-**gindəry*** 'fast etc'.
3.	Hindi	*jəldi* 'fast'	*jəldi-**vəldi*** 'fast etc'.
4.	Marathi	*ləwkər* 'hurriedly'	*ləwkər-**biwkər*** 'hurriedly etc'.
5.	Kannada	*meele* 'above'	*meele-**giile*** 'above and the like'
6.	Bangla	*jore* 'quickly'	*jore-**ţore*** 'quickly etc'.

Table 6: Pronominals.

No.	Language	Word	Echo Word Formation
1.	Tamil	*əvən* 'he'	*əvən-**kivən*** 'he etc'.
2.	Telugu	*ətənu* 'he'	*ətənu-**gitənu*** 'he etc'.
3.	Hindi	*ham* 'we'	*ham-**vam*** 'we etc'.
4.	Marathi	*tula* 'you'	*tula-**bila*** 'to you etc'.
5.	Kannada	*əvən* 'he'	*əvən-**givən*** 'he etc'.
6.	Bangla	*tomra* 'you'	*tomra-**ţomra**/* 'you people etc'.

Appendix 2: Expressives in Himalayan languages (*Panchendriya*)

1. Sense of sound: Acoustic noises

Table 7: Animal noises.

No.	Language	Expressive sound		Meaning of expressive
1.	Khasi	*ŋəŋ-ba* to make echo sound	*ŋəŋ ŋəŋ* EXPR	'Mosquito noise'
2.	Tangkhul	*sip sip*		'Sound of cicada twittering'
3.	Meitei	*ŋiyau ŋiyau* *məu məu* *ci ci*		'Mewing sound of cat' 'Barking sound of dog' 'Crying sound of mouse'
4.	Chakashang	*krü krü*		'Sound of hen calling the newly hatched chicken'
5.	Yimchunger	*cik cik* *triyak triyak*		'Hissing sound of snake' 'Monkey's chattering when about to sleep'
6.	Mizo	*bao? bao?* *crit crit*		'Sound of dogs barking' 'Twittering sound of a house cricket'

Table 8: Noises made by humans.

SL No.	Language	Expressive sound		Meaning of expressive
1.	Khasi	*nok-ba* laugh	*k^he k^he* EXPR	'Giggle'
2.	Tangkhul	*t^hi t^hi* *rok rok*		'Giggle' 'Sound of snoring'
3.	Meitei	*k^hi k^hi* *wa: wa:*		'Giggle' 'Laughter' (heartily)
4.	Yimchunger	*ha ha* *k^həm k^həm*		'Sound of laughing in a silly manner' 'Threatening noise with a serious look'
5.	Chakhashang	*tü tü*		'Sound of mourning or crying due to pain or extreme fear'

Table 9: Noises made by natural phenomenon.

SL No.	Language	Expressive sound	Meaning of expressive
1.	Khasi	*k^{h}om-ba* *grəŋ grəŋ* blow EXPR	'Thundering sound'
2.	Tangkhul	*cək cək* *kuŋ kuŋ*	'Sound of drizzle' 'Sound of thunder'
3.	Meitei	*cro cro* *krəŋ krəŋ* *brek brek*	'Sound of heavy rains' 'Sound of thunder' 'Sound of discontinuous rain accompanied by hail'
4.	Nagamese	*tak tak*	'Sound of rain pattering'
5.	Konyak	*pruk pruk*	'Sound of water flowing in small streams'

Table 10: Noises made by miscellaneous inanimate objects.

SL No.	Language	Expressive sound	Meaning of expressive
1.	Khasi	*lau-ba* *tchrin tchrin* to make sound EXPR 'To jingle'	'Jingle' (sound of anklets, bangles etc.)
2.	Tangkhul	*k^{h}ok k^{h}ok* *ten ten* *šeo šeo*	'Sound of wood chopping' 'Sound of tin or plates falling on the ground' 'Sound of fermented rice bubbling'
3.	Meitei	*p^{h}rak p^{h}rak* *pak pak*	'Sound of bamboo breaking' 'Sound of slapping, kicking'
4.	Yimchunger	*t^{h}rak t^{h}rak*	'Sound of fire cracker, bursting of bamboo when put on fire'
5.	Chakhashang	*kruŋ kruŋ*	'Sound of ringing bell'
6.	Konyak	*p^{h}ruk p^{h}ruk*	'Sound of boiling water'

2. Sense of sight

Table 11: Shimmering aspects of objects.

SL No.	Language	Expressive sound	Meaning of expressive
1.	Khasi	*t^həwan micak* *ŋan-ba* *kup^{h}et kup^{h}et* star shine EXPR 'The stars are twinkling'	'Twinkling'
2.	Tangkhul	*p^{h}ik p^{h}ik* *han han*	'Twinkling' 'Glare' (unbearable to the eyes)
3.	Meitei	*hiŋ hiŋ* *raŋ raŋ*	'Shining of metallic object' 'Brightness seen by a weeping person'
4.	Yimchunger	*rək rək* *nciŋ nciŋ*	'Twinkling, sparkling' 'Shining of metallic objects'
5.	Nagamese	camo camo	'Twinkling or momentary shining'

3. Sense of touch

Table 12: Sense of touch.

SL No.	Language	Expressive sound	Meaning of expressive
1.	Khasi	*tsənap tsənap* *pa-pɔt* EXPR thing 'Sticky thing'	'Sticky'
2.	Tangkhul	*hat hat* *lər lər*	'Feeling of sharpness such as knife, thorn etc.' 'Feeling hard when pressed' (strong muscles, cycle tyres etc.)
3.	Meitei	*pet pet* *k^{h}ək k^{h}ək*	'Soft such as overcooked rice or over ripen fruit' 'Hard feelings of woods, bones, shoes etc.'
4.	Chakhashang	*to to* *pa: pa:* *tü tü*	'To feel hard to chew, break or press' 'Loose or sticky like overcooked rice or over ripen fruit' 'Very sticky' (hard to be removed or peeled off)

SL No.	Language	Expressive sound	Meaning of expressive
5.	Yimchunger	*əyik əyik*	'Very hard, such as under done meat, tough'
		t^hrim t^hrim	'Feeling of rubbing legs, hands etc. or touching only the hair not the skin'
6.	Konyak	*yiŋ yiŋ*	'Feeling of electric shock'
7.	Nagamese	*syap syap*	'Sticky like over cooked rice'

4. Sense of smell

Table 13: Sense of smell.

SL No.	Language	Expressive sound	Meaning of expressive
1.	Khasi	*jiŋ sma* *hek hek* smell EXPR 'Unpleasant smell'	'Smelly' (unpleasant)
2.	Tangkhul	*hik hik*	'Strong smell (good or bad) causing irritation in the nose'
3.	Meitei	*swe swe*	'Intense bad smell such as rotten things or stool'

5. Sense of taste

Table 14: Sense of taste.

SL No.	Language	Expressive sound	Meaning of expressive
1.	Nagamese	*sətə sətə*	'Tangy taste'
2.	Tangkhul	*mat mat*	'Extremely hot (chili hot)'

6. Other senses

Table 15: Other senses.

SL No.	Language	Expressive sound	Meaning of expressive
1.	Khasi	*huri hura*	'Confusion'
2.	Tangkhul	*k^hak k^hak* *tuk tuk*	'Feeling of foreign particle in the eye' 'Beating of heart due to longing or depression'
3.	Meitei	*ciŋ ciŋ* *uru uru* *t^huk t^huk*	'Feeling of pain rhyming with the beating of pulse' 'Nauseating feeling' 'Feeling of depression or longing'

Appendix 3: Expressive of 'manner of walking' in Khasi and Tangkhul Naga (Abbi & Victor 1997)

Khasi (Austroasiatic)

***yaid* (v) 'go, walk, proceed'**

bak'-bak'	'go hurriedly'
biaŋ-biaŋ	'walk continuously'
bran-bran	'go very fast'
brum-brum	'go with heavy steps'
hai-hai	'walk shakingly as if very heavy'
han'-han'	'walk like a duck'
kep'-kep'	'go slowly not sure to keep the equilibrium'
ker'-ker'	'walk trembling'
khne?-khne?	'walk like a lame man'
khniŋ-khniŋ	'walk lamely' (as if very tired or there is a stone in the sole)
khrup' khrup'	'walk quickly and stamping the floor with the steps'
kjik'-kjik'	'walk as if on pins'
knia?-knia?	'walk nicely and willing to speak'
knip'-knip?	'walk with pain'
kor'-kor'	'walk trembling (from sickness)'
kthai-kthai	'walk well-dressed'
kthek'-kthek'	'walk like dancing as on springs'
kui-kui	'move a short but large body'
kynrum'-kynrep'	'go on pouncing'
kyrthek'-kyrthek'	'walk like dancing'
kyntup'-kyntup'	'walk dressing very modestly'
dar-dar	'walk briskly'
der-der	'walk about with flying clothes'
dat' dat'	'walk quickly without turning to left or right'
doy'-doy'	'walk quickly (small boy)'
don'-don'	'walk like a bird or child'
dop'-dop'	'walk like a child, who has just learned to walk'
dot-dot	'move as an old person'
neŋ-neŋ	'walk like an intoxicated'
ner-ner	'move shakingly'
nuŋ-nuŋ	'went on walking'

hir-hir	'go longingly'
yor-yor	'walk slowly from weakness'
jaw-jaw	'go about in poorly wet clothes'
ibeŋ-ibeŋ	'go completely naked'
luŋ-luŋ/leŋ-leŋ	'go in a hurry'
suki-suki don-don	'went out very slowly as an old person'
mIen-mIen	'go about very healthy and muscular'
rymphat'-rymphat'	'go about dirty and poorly dressed'
sak-sak	'walk straight on'
san-san	'daddle, walk as if not sure'
dain-ši-dain	'going ahead successfully'
šey-šey	'walk with long strides'
šen-šen	'go as a drunkard'
šop-šop	'walk with caution'
sar-sar	'go stealthily'
suki lwen'-lwen' (v)	'creep slowly'
tai-tai	'go about very dirty'
ter-ter	'proceed in order'
thaid'-ši-thaid'	'proceed on and on'
then'-then'	'walk steady'
thew-thew	'walk with strong legs'
thiaw-ši-thiaw	'walk uphill with strong legs'
thir-thir	'go quickly'
thnet'-thnet'	'walk as if on the point of falling'
thud'-thud'	'walk as stumbling'
thut'-thut'	'walk tremblingly'
twet'-twet'	'walk too fast'
tub'-tub'	'walk as if not liking it'
tub-pa-tub'	'go slowly'
tuin'-tuin'	'go slowly like an elephant'
wai-wai	'walk weekly'
wey-wey	'go in a zig-zag way'
wet-wet	'go on hurriedly'
wit'-wit'	'walk with many obstacles'
wut'-wut'	'go on hurriedly'

Similarly, in Tangkhul Naga walking is perceived in various subtle ways. The verb for ‘walk/go’ is *kəzət* which follows the manner adverbial expressives of several types. Consider:

Tangkhul Naga (Tibeto-Burman)

***kəzət* ‘to go/walk’**

yaŋyay kəzət	‘to waddle like a child (when walked by grownup people)’
šiŋšiŋ	‘to walk with heavy footstep in a direct manner without stopping or looking about’
t^{h}uŋt^{h}uŋ-	‘to walk heedlessly and laboriously, usually with anger or worriness’
yuryur-	‘to walk in batches at a time’
t^{h}utt^{h}ut-	‘to walk stealthily and slowly’
camcam-	‘to walk blindly and slowly; walk like very old people’
həyhəy- ‘	to walk limpingly’
nutnut-	‘to walk unprogressingly with frequent backward motion, as while forcing to go by pushing or dragging’
wuywuy-	‘to walk waveringly, as when one is drunk’
yəyə-	‘to walk in a leisure way without any purpose’
wakwak-	‘to walk with long strides, especially by tall persons’
hiŋhiŋ-	‘to walk fast with rather long strides’
rutrut-	to walk silently and carefully, usually said of thief or persons with suspicious look’
təytəy-	‘to waddle (by around one-year old children)’
p^{h}utp^{h}ut-	‘to walk very fast (as if getting late for some place to reach)’
namnam-	‘to walk straight and quickly not caring for hindrances’
kuku-	‘to walk tiringly with bowed posture’
həkhək-	‘to walk quickly with light steps’
yokyok-	‘to walk swinging the upper part of the body from back to forth, especially by thin and tall person’
haphap-	‘to walk carelessly without looking for what lies on the surface/ground’
yapyap-	‘to walk uneasily with bowed legs, as when one has got boils in the thighs or buttocks’
rinrin-	‘to walk carefully with hesitation, as on thorny surface’
pəkpək-	‘to walk lightly looking back and forth’
t^{h}əkt^{h}ək-	to walk mincingly’
nennen-	‘to walk slowly with hesitation, as with shyness in front or a crowd or someone’
pikpik-	‘to walk fast and swingingly, as in a crowded street or place’

Gregory D. S. Anderson

Reduplication in the Munda languages

Abstract: Munda languages make extensive use of both total and partial reduplication patterns, total reduplication in combination with infixation or prefixation, and complex reduplication patterns with consonant or vowel replacement. Reduplication is often formally distinct but functionally overlapping in Munda languages. Partial, total and complex reduplication are all attested. Reduplication performs lexical and grammatical functions in Munda and is common in 'expressive' vocabulary that frequently encodes defective/unusual characteristics of events or their participants. Grammatically, reduplication can be found in non-finite verbal functions, as well as in marked voice/valence constructions.

Keywords: Munda languages, Sora, Ho, Gtaʔ, total reduplication, partial reduplication, complex reduplication, expressive formations

1 Introduction and overview

The Munda family represents a diverse group of languages that constitutes the westernmost branch of the Austroasiatic language phylum. Munda languages make extensive use both total and partial reduplication patterns, total reduplication in combination with infixation or prefixation, as well as complex reduplication patterns with consonant or vowel replacement, the latter pattern sometimes called "reduplication with fixed segmentism" (Alderete et al. 1999), "echo formation" in the South Asianist literature, expressive formation in the Southeast Asian linguistic tradition, total reduplication-cum-variation by Stolz et al. (2011), and here called complex substitutive reduplication or complex reduplication with consonant or vowel replacement. Reduplication comes in a range of formally distinct but often functionally overlapping constructions in

Gregory D. S. Anderson: Living Tongues Institute for Endangered Languages & University of South Africa, Department of Linguistics and Modern Languages, Theo van Wijk 9–88, Muckleneuk campus, Preller Street, Pretoria, South Africa. E-mail: gdsa@livingtongues.org

https://doi.org/10.1515/9783110599329-002

Munda languages, whatever the theoretical approach taken to describe this phenomenon.[1]

In this study I focus on Ho, a language from the Kherwarian sub-group of North Munda with over one million speakers, and on two languages from the southern part of the Munda territory, Sora, with around 300,000 speakers, and Gtaʔ the smallest of the Munda languages spoken in southern Odisha by under 5,000 people. Representing the maximal typological spread within the family, across these three languages one finds a range of both formal and functional sub-types of reduplicative formations that typify the Munda languages, as well as various archaic patterns. Data sources for my study include both published materials on the languages as well as unpublished field notes from Living Tongues Institute's Munda Languages Initiative. To the former type belong foremost Deeney's *Ho-English dictionary* (1978) and Ramamurti's (1938) *Sora-English dictionary.*

In Sections 2 and 3, I briefly introduce examples of partial reduplication and total reduplication in the Munda languages Ho, Sora and Gtaʔ. In Section 4, I discuss complex reduplication patterns with consonant or vowel replacement, or both. In Section 5, I turn to the functions of reduplication in the Munda languages, both lexical functions, particularly in the creation of the elaborate 'expressive' formations that typify the languages (5.2), as well as grammaticalized uses of reduplication (5.3) attested in various Munda languages.

2 Partial reduplication

In partial reduplication in Munda languages, the reduplicant represents only a portion of the base, usually the first consonant and vowel, but sometimes, as in Gtaʔ, a leftward, regressive or prefixed copy of just C_1 is the default pattern attested.[2]

1 And there are quite a few competing approaches to be sure. There are optimality and other theoretically oriented approaches, such as Downing (1998), Marantz & Wiltshire (2000), Raimy (2000) or the morphological doubling theory of Inkelas & Zoll (2005), creolistic approaches to reduplication, like Bakker & Parkvall (2005) and also typological approaches to reduplication and related phenomena, e.g., Moravcsik (1978), Rubino (2005), Hurch (2005); Wälchli (2005, 2007), Hurch & Mattes (2007, 2009), Rozhanskij (2011), Schwaiger (2011), Aboh et al. (2012) or Stolz et al. (2015).

2 Gtaʔ sources refer to the sources used in my forthcoming dictionary (Anderson in preparation-a, which includes unpublished field notes from the 1960s by Mahapatra and Zide), text

(1) Gtaʔ

t.tæʔ=riaʔ	'water plants'	[M89]
t.tua	'rub, massage'	[M89]
c.cu	'fruit'	[Anderson in prep-a]
c.cog	'to string beads'	[M89]
p.pa	'to look searchingly'	[M89]
p.paʔ	'to brush away, flick'	[M89]
p.poʔ=gi	'pick tooth'	[M89]
r.riʔ	'eat flesh from bones'	[M89]
m.mweʔ	'extract & expel seeds'	[M89]
n.naʔleɽia[ʔ]	'lightning'	[M89]
n.naʔloʔco	'steps'	[M89]
l.li	'creeper'	[Anderson in prep-a]
l.lir	'to cut thin long strips'	[M89]
l.locoŋ	'to beg'	[M89]
l.loʔriaʔ	'to drizzle'	[M89]
l.luæ	'pubic hair'	[M89]
l.lwar	'leak through a hole'	[M89]
k.kwi	'wrap'	[M89]
k.koʔ	'to cough'	[M89]
k.kaʔ	'donkey'	[M89]
ɟ.ɟo(ʔ)	'put in, throw into'	[M89]
h.hia	'festival'	[M89]
g.gaʔ	'crow'	[Anderson in prep-a]
g.gusweʔ	'whistle'	[M89]
g.go	'go hunting in groups'	[M89]
g.gæ	'chew, masticate'	[M89]
g.gok	'hatch egg, peck, graze'	[M89]
g.gu=riaʔ	'cloud(s)'	[M89]
ɖ.ɖo	'feel desire, attraction'	[Anderson in prep-a]
b.ba	'pat'	[Anderson in prep-a]
b.buʔ	'suck'	[M89]

The only stems that show a full syllable copy (C_1V_1-) in a reduplicated form in Gtaʔ begin with the following sequences: *gu-* and *gi-*.

collection (Anderson in preparation-b) and grammar (Anderson in preparation-c). These include, in addition to my field notes, an Oriya-language publication by Mahapatra et al. (1989) [M89], and the brief lexicon in Ashirvadam's (1992) PhD dissertation [A92].

(2) Gta?

gu.gogo	'give birth (animal)'	[A92]
gu.g[u]ra	'ring dove'	[M89]
gu.guar	'scrape out'	[M89]
gu.gua?	'chop vegetables'	[M89]
gi.giŋ=ḍia?	'gargle and spit out'	[M89]

Note that almost all the lexicalized forms with partial reduplication in Gta? in (1) are either inherently or canonically pluractional (verbs) or appear canonically in groups (nouns). But partial reduplication has numerous morpholexical functions in Gta?. Among these are to be included infinitives and participles of verbs, e.g., *ggo=sa* 'hunting story', *ha.hæjg* 'excrement, shit(ting), to shit', or *ha.ha?* 'to bite', and the formation of certain nouns like *hha* 'arrow' (Anderson in preparation-a, -b, -c). In Plains Gta?, infinitive complements undergo partial reduplication if monosyllabic, but not if disyllabic, as in serialized structures, compare (3a) and (3b).

(3) a. Plains Gta? vs.

næŋ sela ttur ũ-we=e
I girl REDPL:search 1-go=FUT
'I will go search for a girl to marry'

b. Plains Gta?

næŋ e+tur ŋ-ke=e
I go+search 1-ATT=FUT
'I will try and go search for her'

Partial reduplication has also been grammaticalized in a number of other southern Munda languages. For example, the imperfective/progressive auxiliary *ḍen* requires monosyllabic verb stems to be reduplicated in Remo and Gutob (Anderson in preparation-d).

(4) Remo

niŋ nsuṛa? susum ḍen-t-iŋ
I banana REDPL~eat AUX-NPST-1
'I am eating a banana' [Field notes]

(5) a. Gutob

o-niŋ dunuŋ-laj bebe?-pen ura?
OBJ-I run-PURP REDPL:CAUS-2PL NEG.COP
'you didn't make me run' [Field notes]

b. Gutob

loko=nen loko=nen majrama= ḍəḍen=nen ura?
man=PL man=PL hit REDPL~AUX=3PL NEG.COP
'the men will not hit each other' [Field notes]

C_1V_1 partial reduplication is relatively rare in Ho lexemes, and the forms sometimes alternate with un-reduplicated simplicia. Note however that iterative bases in Kherwarian languages as a whole take a partially reduplicated stem allomorph that is generally of the shape C_1V_1-, e.g., Santali *dal* 'beat' *dadal* 'repeatedly beat'.

(6) Ho

ba.bata	'to itch'
ka.kaṛi ~ kaṛi	'to chase'

3 Total reduplication

In total reduplication, there is a full copy of all segments, that is, full identity between the base and the reduplicant (Stolz et al. 2011). Within total reduplication itself, nothing *per se* in the Munda languages dictates whether this pattern is leftward/regressive/prefixal or rightward/progressive/suffixal. However, in complex substitutive reduplication or reduplication with consonant and/or vowel replacement (also known as 'echo' formation), such patterns can be discerned when comparing unreduplicated/unaltered bases with their reduplicated counterparts, when they exist. So, in Sora, we can assume that the default situation for the pattern in complex reduplication appears to be Base-Reduplicant, i.e., it is a rightward, progressive or suffixed copy pattern. In 'echo' forms in Ho on the other hand, the directionality of the copy appears to be both leftward/regressive/prefixed and rightward/progressive/suffixed. In the following overwrite formation, /ẽ/ is clearly etymological, and thus, the form *cã.cẽ* must be Reduplicant-Base, while forms such as the one in (20), *nili?.**n**i**p**i?*, where the -*l*- is etymological and -*p*- the replacing consonant, the form must rather be Base-Reduplicant.

(7) Ho

cã.cẽ	'sound of many baby crying at once'
cẽ.cẽ	'sound of one baby crying repeatedly'

In (7), the totally reduplicated form is iconic of repeated action, while the related form with over-write of [ẽ] with [ã] encodes here rather distributed pluractionality of many subjects acting at once. If the echo form with complex substitutive reduplication is taken as the formal model in this case, then total reduplication could also be considered to be variably regressive/leftward/prefixed (Rozhanskij 2011) in Ho but exclusively progressive/rightward/suffixed in Sora. This is noteworthy not only because the two languages differ in this regard but overall show the same replacing patterns in vowels and many of the same consonants, see Section 4 below, but also because as a rule, Ho has historically eliminated almost all derivational prefixes in favor of nearly exclusive suffixation (and a handful of infixes), but Sora still uses many archaic inflectional and derivational prefixes with parallels in other non-Munda Austroasiatic languages.

Total reduplication is common in both Sora and Ho. Many verbs, nouns, adjectives, adverbials, and ideophones show total reduplication.

(8) Sora

riŋge-n	*boŋboŋ=loge*	*tid-t-ai*
wind-NSF	whizzing=ly	blow-NPST-1/CLOC

'the wind blows whizzingly towards me' [Ramamurti 1938: 63]

Reduplicants may be up to three syllables in length in both Sora (9) and Ho (10).

(9) Sora

bad.bad	'to coddle'; 'to unfurl'
aɟid.aɟid	'little by little'
aɟiger.aɟiger	'splashing of water'

(10) Ho

bo:.bo:	'sound made by snake pushing along ground' (+ verb of motion)
bal.bal	'to sweat'
biḍ.biḍ	'on tiptoes'
biɟil.biɟil	'flash of light reflected off disturbed water, lightning flash'
biṭil.biṭil	'repeatedly move body up and down'
bijuŋ.bijuŋ	'feel a spinning sensation in head', 'giddy'
banḍul.banḍul	'with bushy tail wagging'
beteʔ.beteʔ	'tear, cut into small pieces'
beṭoʔ.beṭoʔ	'move across water in undulating motion' (of water skidder insect)'
bokoṛo.bokoṛo	'sound of boiling water bubbling'

Unlike its sister languages, total reduplication is not as common in Gtaʔ, but some forms are found that allow up to a $C_1C_2VC_3$ syllable or even two syllables to be copied.

(11) Gtaʔ

iici.iici	'make faces at, mock'	[M89]
bluŋ.bluŋ	'sisters to each other'	[Anderson in preparation-a]
biŋ.biŋ ~ bim.biŋ	'earthworm'	[M89 ~ A92]
mæʔ.mæʔ	'which ones'	[Anderson in preparation-a]

4 Complex reduplication with consonant and/or vowel replacement

Complex substitutive reduplication, i.e., reduplication with replacement or overwrite of consonants, i.e., with consonant and/or vowel replacement, is widely attested in both Ho and Sora, with many formal sub-types, but it is not very common in Gtaʔ. Complex reduplication patterns serve to express a wide range of functions in both Sora and Ho, and it is likely that complex reduplication was a feature of the Proto-Munda ancestral language as well. Indeed, Austroasiatic languages in general have an extensive set of affective vocabulary formed by complex reduplication, the so-called 'expressive' vocabulary (Diffloth 1979). Thus, postulating such a system for Proto-Austroasiatic also seems justified. Complex reduplication formations therefore have a long history in the Munda languages and constitute an integral and emblematic part of the languages as used by their speakers.

Complex substitutive reduplication represents a co-lexicalization of base and reduplicants that are only partially phonologically related to the base, with a pre-determined and fixed replacement in the reduplicant of various consonants and vowels of the base. The best-known cases in the literature on 'echo'-formation involve single consonants or substitutive syllables as the dominant or even unique option for 'echo'-formation, as in Hindi *v-* or Tamil *ki-* (12), see Abbi (1992), Singh (2005).

(12) a. Hindi [Singh 2005: 265–266]
*chaay.**v**aay* *Narendra.**V**arendra*
'tea, etc.' 'Narendra and similar undesirables'
b. Tamil [Parimalagantham 2009: 26]
*kallu.**ki**llu* 'stone, etc.' *maram.**ki**ram* 'tree, etc.'

A characteristic feature of the South Asian linguistic area, such formations are found in a wide range of Indo-Aryan languages, e.g., Panjabi (Ahmad 1963), Marathi (Apte 1968), Dakhini (Mustafa 1979) or Hindi (Singh 1982, 2005), and in Dravidian languages as a whole (Emeneau 1938, 1969), e.g., Telugu (Bhaskararao 1977, Selvam 1988) or Tamil (Wiltshire 1999).

In Gtaʔ (Anderson in preparation-b, -c), there are co-lexicalized forms where the first word and second word bear some resemblance to each other phonologically. In these, the first word, which can usually appear alone syntactically in the same or related meaning, is paired with the second, which sometimes cannot. The first syllable is copied but the second syllable is varied in one such pattern involving such paired sets of disyllabic forms:

(13) Gta?

baʔ_lir-_***baʔ***_taʔ_	'converse'	[Anderson in preparation-a]
seʔ_gwaʔ+_***seʔ***_riŋ_	'threaten'	[M89]
bo_seʔ+_***bo***_to_	'make offerings in name of spirits'	[M89]
ur_guʔ+_***ur***_laŋ_	'to pull'	[M89]
saʔ_kur+_***saʔ***_wæ_	'millet pudding'	[M89]

Similar formations are found in Sora too:

(14) Sora

sə_tid+_***sə***_boi_	'smarting sensation'

A second pattern is attested where the second syllable is copied but the initial syllable varies:[3]

(15) Gta?

kɽe***soʔ***_+hnaʔ_***so(ʔ)***	'hungry'	[Anderson in preparation-a]
gguɽ***iaʔ***_-læŋ_***ɖiaʔ***	'cloud'	[M89]

Sometimes, what is copied in this type of pattern is longer than a single syllable. Forms with two (16) or up to three copied syllables (17) can be found in Sora.

(16) Sora

_g_***əloi***_-ɟ_***əloi***	'to slip'

(17) Sora

_ə_***da:ɟado:ŋ***_-go_***da:ɟa-doŋ***	'swear'

Note the following case in Ho, where what is copied is a coda of the first syllable and the entire second syllable, which is a unit, like in (16) in Sora above, that has no other phonological reality in the language.

(18) Ho

_a_***rkam***_-tu_***rkam***	'in any way at all, e.g., work spiritlessly'

More complex alternations can be found, as for example in the following forms that keep the onset and coda the same but alternate the medial element in Gta? (19).

3 The difference between the two patterns appears to be as follows: in the second pattern as in (15), the two forms have identical combining forms together with distinct roots, i.e. the combining form is copied. In the first type (13), a root is copied, and different second elements are used. However some of these elements are clearly combining forms of nouns, e.g., *=ta(ʔ)* 'mouth' in *baʔ=ta(ʔ)*, the 'echo' form of *baʔlir*, in other cases this is not the case, and the alternating forms appear to be verb stems themselves (e.g., *guʔ* means 'pull' by itself).

(19) Gtaʔ

***s**wiŋ+**s**aŋ*	‘quickly and quietly’	[Anderson in preparation-a]
gro+gco	‘government official’	[Anderson in preparation-a]

Medial consonant replacement is found in other Munda languages as well in complex substitutive reduplication, e.g., in the Kherwarian language Ho, where *-p-* in the reduplicant can replace base-medial *-l-*:

(20) Ho

*nili**ʔ**.ni**p**iʔ* ‘intense heat’

Again, the copied elements can include up to three syllables in Sora. In Sora, unlike ‘regular’ complex substitutive reduplication where labials predominate as the replacing consonant in the echo form/reduplicant, rather the medial replacing consonant *-l-* is a common choice, replacing *-r-* or *-b-* (21).

(21) Sora

ə**r**ədi:ɟa:-ə**l**əti:ɟa:	‘even a bit (wet or dry)’
da**b**umloge+da**l**umloge	‘sound caused by striking’

More complex alternations of this ‘internal replacement’ type are also attested in a small number of forms in Sora ‘echo’ words as well.

(22) Sora

ano**puŋ**lai+ano**m**lai ‘saying’

In addition to actual complex substitutive reduplication, where there is partial similarity between base and reduplicant or copy, such that we can speak of cases of reduplication, there are also instances of simple co-lexicalization where the associated echo-like form is *not* phonologically related to the first element in such a co-compound, and occurs only with the specific ‘base’ and without independent meaning in the language as in the following Gtaʔ forms.

(23) Gtaʔ

ugboʔ+canḍriaʔ	‘hair’	[M89]
briŋ+taʔma	‘bend one’s body’	[M89]

Many complex reduplication patterns with replacement are found in Sora and Ho. One finds patterns of complex reduplication with consonant replacement in the reduplicant, vowel replacement in the reduplicant and consonant + vowel replacement in the reduplicant.

4.1 Complex reduplication with consonant replacement

In Sora, complex reduplication with replacement of consonants in the reduplicant is widespread. In (24), I list some of the replacement patterns attested in the data set:

(24) Sora

Base	Copy (reduplicant)	Base	Copy (reduplicant)
Ø-	*b-, s-*	*p-*	*m-*
g-	*b-, s-*	*k-*	*m-*
l-		*b-*	*t-m-*
ɲ-	*p-*	*m-*	*p-*
r-		*b-, p-*	*s- m-, p-*

As a whole in Sora, if the base starts in a voiceless obstruent, the replacing consonant in the reduplicant is always *m-* (25), which is otherwise only found in Sora as a replacing consonant with bases that start with *s-* in a handful of examples (26).

(25) Sora
*taŋlij.**m**aŋlij* 'cattle'
*tonai.**m**onai* 'witchcraft'

(26) Sora
*so:la:n.**mo**:la:n* 'evil spirits'
*so:ra:.**mo**:ra:* 'Sora and other such people'

Sora bases that begin in *s-* may also be replaced in complex reduplication with *p-*.

(27) Sora
*sədirga:mle.**p**ədirga:mle* 'thin (of porridge)'

For bases starting in vowels or *g-*, the added or replacing initial consonants in the reduplicant can be *b-* or *s-* only.

(28) Sora
gelo:ŋ.belo:ŋ 'hurried manner of gobbling down food'
***geluŋ.b**oluŋ* 'confused'

Sora bases that start in *l-* can only be replaced by *b-* in the reduplicant, while *r-*initial bases in Sora can be replaced in complex reduplication by *b-* or *p-*.

(29) Sora
*lijeŋ.**b**ojeŋ* 'in tattered condition'

(30) Sora
__r__ided.__b__oded ‘in a loose, rickety condition’, ‘limp, lame’

Bases that start in nasals in Sora tend to replace the initial nasal in the base with *p-* in the reduplicant.

(31) Sora
__ɲ__a:ɟe:ŋte.__p__aɟɟe:ŋte ‘toddle like baby walking’

Thus, the only consonants found in reduplicants in Sora are the labials *b-*, *m-*, *p-*, except with a small number of vowel- or *g*-initial words where that consonant can be *s-* and one example where *ɟ* in the base alternates with *g* in the reduplicant. In Sora’s distant sister language Ho, which belongs to the Kherwarian North Munda subgroup (32) and spoken in southern Jharkhand and northern Odisha, these three labial sounds also form the overwhelming majority of the instances of consonant replacement in complex reduplication. However, unlike in Sora, *m-* is by far the most common replacing consonant in the reduplicant in complex reduplication of this formal type in Ho.

(32) Ho

Base	Copy (reduplicant)	Base	Copy (reduplicant)
ʈ-	*m-*	*h-*	*b-, p-, (t-)*
b-	*m-*	*s-*	*b-, (p-)*
c-	*m-, b-*	*Ø-*	*m-, b-, p-, c-, s-*
ɖ-	*m-*	*r-*	*m-, b-, p-, c-, (g-)*
d-	*m-, p-*	*l-*	*m-, b-, p-, c-, (k-)*
g-	*m-, (c-)*	*k-*	*m-, b- (d-, ɖ-)*
ɟ-	*m-, p-, s-*	*(n-*	*c-, d-)*
t-	*m-, b-*		

(33) Ho
aʈa.__m__aʈa ‘very thick of jungle’
ikiɖ.__m__ikiɖ ‘walk with arms swinging proudly and ostentatiously’
uʈi.__m__uʈi ‘full of knots’ (tree) ‘very thin and bony’ (person)
__c__awor.__m__awor ‘confused din of many people speaking together’
__c__enɖeloŋ.__m__enɖeloŋ ‘be undersized and thin’ (veggies, fruit)
__d__ela.__m__ela ‘have large stomach’ (small child)
__d__oɖ.__m__oɖ ‘stammer, stutter’
__d__aʈuʔ.__m__aʈuʔ ‘go along w/ head bobbing up and down’
__g__oe.__m__oe ‘wilt, bend over’
__ɟ__olo.__m__olo ‘clear and sparkling’

*koroeʔ.**m**oṭoeʔ*	'to grumble'
*ṭaṭub.**m**aṭub*	'walk along unsteadily due to weakness at knees'
*tanḍaʔ.**m**anḍaʔ*	'walk with legs spread far apart'

The labial obstruents *p-* and *b-* are the next most common replacing consonants in the reduplicant in Ho complex reduplication patterns.

(34) Ho

*aka.**b**aka*	'be dumbfounded, be confused'
*ceṭe.**b**eṭe*	'noise made by rice beer as it ferments'
*ciːri.**b**iːri*	'smarting sensation on skin after scrubbing and soaping on cold day'
*harom.**b**arom*	'collection of contributions for village sacrifice'
*hae.**b**ae*	'suffer hardships'
*kaeḍ.**b**aeḍ*	'have disturbed feeling in stomach'
*loṛo.**b**oṛo*	'speak almost unintelligibly'
*roːṛo.**b**oːṛo*	'give off smell'
*sago.**b**ago*	'low rumbling breathing of weak person'
*segel.**b**egel*	'scatter in disorder'

(35) Ho

*dasi.**p**asi*	'house servant, slaves'
*dukur.**p**ukur*	'be anxious with concern or fear'
*leṭe.**p**eṭe*	'pasty, form a paste'
*hiriʔ.**p**iriʔ*	'scatter in a disorderly fashion'
*roṭoeʔ.**p**oṭoeʔ*	'weak and thin due to sickness'

A handful of forms with liquids in initial position in the base (*l-* or *r-*) use *c-* in the reduplicant.

(36) Ho

*raga.**c**oga*	'be very rough and uneven'

Three forms in the corpus use *s-* as the initial replacing consonant in the reduplicant. All other consonants occur only once or twice.

(37) Ho

*uti.**s**uti*	'in one's full senses'

Ho also has a relatively common word-*internal* replacing pattern in complex reduplication where one finds an alternation between *-r-* in the base and *-ṭ-* in the reduplicant.

(38) Ho

Base	Reduplicant
-r-	*-ṱ-*

(39) Ho

ɟirui.ɟiṱui	'hanging in tatters'
ɟuru.ɟuṱu	'loaded down with many things such that walking is difficult'
sere.seṱe	'noise of porcupine tail moving'

This internal medial consonant replacement pattern can also co-occur with initial consonant replacement (*k* > *m*), and internal vowel replacement (*e* > *u*), or both (40).

(40) Ho

kere.mere ~ kere.muru ~ kere.muṱu	'quickly, in a rush'
koroeʔ.moṱoeʔ	'to grumble'
koroʔ.moṱoʔ	'cause painful congested sensation in nose'
kuru.muṱu	'do something in excited haste'

4.2 Complex reduplication with vowel replacement

In addition to the patterns of consonant replacement seen in Section 4.1 above, both Sora and Ho show complex reduplication patterns with vowel replacement as well. As seen in (41), the default vowel in the reduplicant is (-)*a*- in Sora, which can replace all the other base vowels.

(41) Sora

Base	Copy (reduplicant)
o	*a*
u	*a*
e	*a*
i	*a*
(ə)	*(a)*

This restriction to *a* as the replacing vowel in the reduplicant is particularly pronounced with respect to the bases that have *o*, *e*, and *i*, which really only alternate with *a* in reduplicants, except in one or two examples each.

(42) Sora

*ba:r**o**:i.ba:r**a**:i*	'rumbling (as of intestines)
*r**o**b.r**a**b*	'to rustle'

po?de.pa?de=ge	'in coils, in folds'
ɲidur.ɲadur=ge	'glittering (like cat's of eyes at night)'
sir.sa:r	'be split', 'sprinkle, spill'
me:ŋ.ma:ŋ ~ me:ŋ.me:ŋ	'humming of bees, insects'
reb.ra:b	'sound of tree/branch cracking and falling'
rum.ram	'tread, trot, prance on horse'
pakuŋge.pakeŋge	'in a rage, furiously'

The vowel *u* in the base also alternates with *a* primarily, except in four words where *u* in the base alternates with *e* in the reduplicant.

(43) Sora

dub.deb	'disgusting'

The high front vowel *i* never occurs in the reduplicant in Sora, and *e*, *o*, and *u* are marginal. These minor patterns include the following:

(44) Sora

Base	Copy (reduplicant)
o	*(e)*
u	*(e, o)*
e	*(o, u)*
i	*(o, e)*
(ə)	*(o)*[4]

Note that when the vowel *a* occurs in the base, it may alternate with *u* or *o* in the reduplicant in Sora complex substitutive reduplication.

In Ho, *a* is also the main vowel in the reduplicant that replaces a range of vowels in the base. In a few examples, base vowel *a* (and to a lesser extent *o*) may be replaced by *i* in Ho reduplicants as well (45).

(45) Ho

Base	Copy (reduplicant)
u, o, e, i	*a*
a, o	*i*

In a small number of examples one finds replacing vowel *u* with base vowel *a*, and replacing vowel *e*, corresponding to base vowels in *o*, *u*, *i*, and *a* (46), but

4 The vowel *ə* is not overly common in bases in Sora and only marginally participates in complex reduplication with vowel replacement, but the few examples show either *a* or *o* in the reduplicant.

only one to two examples are attested for each. Note that *o* is never a replacing vowel in Ho reduplicants.

(46) Ho

Base	Copy (reduplicant)
(a)	*(u)*
(i, u, o, a)	*(e)*
*	**o*

(47) Ho

*baŋ.**boŋ***	'make a gaping hole in'
*chapel.ch**o**p**o**l*	'splashing sound made by walking through several inches of water'
*ba:ḍ.**be:ḍ***	'cries of many sheep and goats'
*baṛa:ḍ.**beṛe:ḍ***	'cry of many goats'
*pala?.**pili?***	'flash on and off'
*par.**pir***	'scatter in all directions'

Like Sora, one, two or three syllable bases can be found in patterns with complex reduplication with vowel replacement in Ho.

(48) Ho

*ba:.**bu:***	'look here and there when lost or looking for something'
aka.uku	'run in an ungainly due to fatness'
ɟanḍala.ɟunḍulu	'thin'

However, only the first vowel alone can also be replaced and the second one maintained in certain disyllabic bases in Ho:

(49) Ho

*la:ka.**lu:ka***	'bulging out in places'

While in others, it is rather the second syllable only where the vowel is replaced, not the first one:

(50) Ho

*bala.**balu***	'mad, stupid, delirious', 'confuse someone very badly'

Bases with the diphthongs *ui* (51) or *oe* (52) both are replaced by *ae* in the reduplicant in complex reduplication in Ho.

(51) Ho

chae?.chui?	'sound of rats/mice', 'sound of whipping'
gaelaŋ.guiluŋ	'move around slowly, mope'

	hae.hui	‘multiple birds’ wings swishing in air’
	kaɽaeʔ.kuɽuiʔ	‘form clods that don’t break down properly when ploughing (soil)’
(52)	Ho	
	karaeʔ.koroeʔ	‘noise made clearing throat’
	rapaeʔ.ropoeʔ	‘uneasy feeling at onset of sickness’
	ʈaeʔ.ʈoeʔ	‘die off quickly one after other’

Note also the complex substitutive pattern of *o* … *o* > *i* … *a* is found with a number of bases in Ho.

(53)	Ho	
	hitaː.hotoː	‘water at mouth when seeing something tasty’
	kidar.kodor	‘be long and wavy’ (cock tail, bird plume)
	lidab.lodob	‘grow full leaves in profusion’
	diŋar.doŋor	‘bright red’

Of course, complex reduplication with consonant replacement (or here addition as the base is vowel-initial), plus vowel replacement is also attested in Ho.

(54)	Ho	
	aʈa.puʈu	‘covered with dense undergrowth, and difficult to pass through’

Complex reduplication with vowel replacement is clearly an old pattern in Austroasiatic. Khasi (Rabel 1976), a distantly related Austroasiatic language spoken in Meghalaya, India, also shows a clear preference to *a* in the reduplicant, regardless of the base vowel in complex reduplication with vowel replacement.

(55)	Khasi	
	Base	Copy
	u	*a*
	i	*a*
	-a	*-i*

Note that in Sora there are a few bases with an initial *l-* that have complex double reduplication pattern, which combines partial Cə- reduplication with either total reduplication (56) or complex reduplication with vowel replacement (57).

(56)	Sora	
	lə-laːm-laːm-ən	‘trellis’
	lə-luː-luː	‘to serve as means of taking rest’

(57) Sora

lə-luɟ-loɟ	'remove outer layer'	

Unlike its sister languages Sora and Ho, complex reduplication with vowel replacement is relatively uncommon in Gtaʔ (58).

(58) Gtaʔ

gi.ge	'to scratch'	[A92]
ɟi.ɟar	'alight'	[A92]

5 Functional domains of reduplication in Munda languages

Reduplication in the Munda languages performs a number of different functions. Some of these are derivational or lexical (5.1). A subset of lexical functions of reduplication in Munda languages includes the highly developed and functionally nuanced system of expressive formations (5.2). Still other functions of reduplication in Munda belong to the domain of grammar (5.3).

5.1 Derivational/lexical functions or meanings

Sora has a pattern of total reduplication plus infixation of either *-ər-* (60) or *-ən-* (61) in the base, and in many cases also appears together with the so-called 'noun' suffix in *-(ə)n* of obscure function (Gomango 2015, Anderson & Harrison 2008, Starosta 1992) that, if present, follows the reduplicant. This yields a pattern as in (59):

(59) CVC > C/-*ər*-/VC-CVC[-*ən*] or C/-*ən*-/VC-CVC[-*ən*].

(60) *-r-* infixation in Sora

*d**ər**ai.daɟ-ən*	'stairs, ladder' < *dai/daɟ*
*d**ər**ul.dul-ən*	'ambush' < *dul*
*g**ər**am.ga:m*	'meaning' < *ga:m*
*g**ər**iɟ.giɟ-ən*	'window' < *giɟ*
*g**ər**ʊd.gʊda:-n*	'basket for bailing water' < *gʊd*
*g**ər**ob.go:b*	'seat' < *go:b*
*p**ər**ad.pad*	'raft', 'meat-skewer' < *pad*
*s**ər**u:.su:-n*	'battle field', 'place of fighting' < *su:*
*t**ər**ub.tub-ən*	'expedient' < *tub*

*d**əri**:.di:*	'vertically' < *di:*
*g**əru**.guden-ən*	'invitation' < *gu*
*d**əruŋ**.duŋjuŋ-ən*	'rising sun' < *duŋ*

(61) -*n*- infixation in Sora

*d**əni**:.di:-n*	'counting, number'
*d**əniŋ**.diŋ-ən*	'dragging, attraction'
*g**əna**:.ga:-n*	'food, drink'
*g**ənu**:.gu:-n*	'seedling, what is sown'
*l**əno**:.lo:-n*	'hoeing'
*m**əne**l.mel-ən*	'examination, inspection'
*r**ənu**:.ru:-n*	'an ornament'
*s**ənu**b.su:-n*	'wooden pot planted in ground'
*t**əno**:l.to:l-ən*	'rope to tie cow to post'
*r**əni**d.[rid]ən*	'powder, flour'
*t**əno**ɟ.[toɟ]-ən*	'stake, wager'; 'sorcery'
*s**əno**i.soi=toŋ-ən*	'antics, kind of dance'
*s**əno**:.so:si:-n*	'children's game type'
*g**ənu**.gu=mar*	'summoner'
*k**ənu**d.kud*	'bear a child'
*t**əno**l.tol-ən*	'marry, live w/ man'

Note that these forms need not be nouns and may function syntactically as verbs in Sora, and indeed may even take undergoer agreement/indexing in polyvalent predicates.

(62) a. Sora
*d**əne**:de-l-am*
wait/NOM/wait-PST-2UND
'I waited for you'

b. Sora
amme:le dede:-l-am
you-for REDPL.wait-PST-2UND
'I waited for you'

In both Gtaʔ (63) and Kharia (64), a number of common nouns appear in a form with partial reduplication and the -*n*- infix. Many of these resulting forms have instrumental semantics or express other derived nominal concepts that transparently relate to primary verbal bases.

(63) Gtaʔ

*g-**n**-ga(ʔ)*	'bundle'	[M89]
*g-**n**-gæ*	'jaw'	[M89]
*g-**n**-giŋ*	'cheek'	[M89]

(64) Kharia [Abbi 1992: 120]
ɟiniɟib 'taboo relating to touching food' (< *ɟib* 'touch')

Also, co-compounds or co-lexicalized echo-like formations are not uncommon with *-n-* infixation in Plains Gtaʔ.

(65) Plains Gtaʔ

bnog=*sæŋ*+***bnog***=*ɟar*	'summer, hot season'	[M89]
pna=*ta*+***pna***=*li*	'harvest time'	[M89]
bnaʔ=*lir*+***bnaʔ***=*ta*	'conversation'	[M89]

One common type of expressive semantics associated with reduplication in Munda languages is intensity, especially with color terms, as in Ho (66).

(66) Ho

gul.gul	'having deep dark color'
che:ra/me:ra	'really beautiful'
diŋar.daŋar	'red, blood colored'
diŋar.doŋor	'bright red'

Sora has both intensified color forms, as well as reduplicated simplex forms like 'yellow'.

(67) Sora

pem.peŋ	'very dark'
saŋ.saŋ	'yellow'

A number of modifiers are reduplicated in the various languages of the Munda family. For example, partial reduplication (C-) is found in Gtaʔ (68). Typically, total or complex reduplication mark such 'expressive' meanings in other Munda languages, see Section 5.2 below.

(68) Gtaʔ

m.moʔræg	'ugly bad-smelling person'	[Anderson in preparation-a]
t.taŋ	'dark'	[M89]
s.sig	'cold'	[M89]

In Sora (69) and Ho (70), one finds total reduplication or complex reduplication with vowel or consonant replacement; such forms can of course appear in larger derivational complexes or in phrases as well.

(69) Sora

dub.deb	'disgusting'
pa:.pa:=u:=boɟ-ən	'woman w/ hair parted in middle'
pa:-.pa:=ʔur-ən	'split bamboo'
da.da=dəm	'firm, tight'

(70) Ho
a: chol.chol 'long beak'
huɽiɲ.muɽiɲ 'small'

Reduplicated adverbial forms are common in Sora. Total reduplication is found in a number of these reduplicated adverbial formations in Sora.

(71) Sora
tuŋ.tuŋ 'extremely'
tetten.tetten 'now and then'
saletten.saletten 'then & there, every now and then'
togəl.togəl 'during the night'

In some such forms in Sora, complex reduplication with vowel replacement can be found as well.

(72) Sora
sir.sa:rloge 'in spray'

More complex formations such as total reduplication + *r*-infixation characterize a handful of reduplicated adverbial forms in Sora.

(73) Sora
*də**ri**:.di:* 'vertically'

Like Sora, reduplicated adverbial forms in Ho can either show total reduplication (74) or complex reduplication with vowel replacement (75).

(74) Ho
chirgal.chirgal 'cautiously'

(75) Ho
daba:.doba: 'in a large group together'

In Sora, a number of small, crawling creatures are formally lexically reduplicated simplicia that do not (typically) occur in non-reduplicated forms.

(76) Sora
kən-di.di:=bud 'millipede'
bub.bud-ən ~ bud.bud-ən 'insect, worm'
gar.gar=bud-ən 'borer insect'
tut.tudia-n 'inch worm'
tut.tum 'bloodsucker'

A number of tree names are also lexically reduplicated simplex forms.[5]

(77) Sora

rar.ra:d=ʔa:bim=ne:b-ən	'achyranthus aspera'
ro.:ro:=ne:b-ən	'tree sp.'
ruŋ.ruŋ-ən	'tree sp.'
kur.kuri=ne:b-ən	'tree sp.'
par.par=ne:b-ən	'tree sp.'

5.2 'Expressive' formations and expressive semantics

Expressive reduplication is a hallmark of Munda language structure, as it is for many related languages within Austroasiatic (Rabel 1976, Nagaraja 1984 [Khasi]; Benjamin 1976 [Temiar]; Diffloth 1976 [Semai]) and indeed the Austroasiatic family as a whole (Diffloth 1988). Both total reduplication and complex reduplication with consonant or vowel replacement can be used to create an enormous range of expressive forms in both Sora and Ho. While a larger data set analyzed systematically (which is not yet possible) may yield other results, there are no significant trends across the different languages that suggest any formal or functional specialization of specification of the different patterns, only individual language-specific cases mentioned below, such as total reduplication in motion-related meanings in Sora and defective speech correlated to complex (substitutive) reduplication in Ho.

5.2.1 Unusual, excessive, deformed

Among the most typical functions of expressives in South Asian languages is to express characteristics that are somehow negative, as in the following Sora words:

(78) Sora

kokede.kakode	'crooked, curved'
puŋ.puŋ	'swollen, bloated'
rided.boded	'in a loose, rickety condition', 'limp, lame'
sos.sor	'rough'
(ə)rige.rige	'having stripes of different colors'

5 Note *=ne:b* is the combining form for 'tree'.

po:.po:=mu	'having sunken nose'
piŋ.piŋ ~ peŋ.peŋ	'be cracked', 'have chinks'
kal.kal	'difficult, inconvenient'
boŋkode.baŋkode	'crooked, awkward'

Ho too has a number of expressive formations that are reduplicated and which have meanings in this same negative or pejorative sense.

(79) Ho

*reṭe.**p**eṭe*	'be contorted' (rope)
raja:.ruju:	'be thin with visibly enlarged spleen'
raga.choga	'be very rough and uneven'
lam[ar].lum[ur]	'gluttonous'
beŋga.biŋgi	'parallel running stripes of alternating colors'

5.2.2 Onomatopoeia, ideophones

Particularly common in reduplicated forms are onomatopoeia and other ideophones reflecting, for example, the ambient acoustic environment in both Ho and Sora. Note that while complex reduplication is found in most but not all examples of this type of form in Ho, the reverse is true of Sora, where rather total reduplication is dominant.

(80) Ho

sai.sui	'swishing sound of raptor plucking up prey'
seṭe.beṭe	'noise of silkworm building cocoon'
ṭal.ṭul	'sound of corn popping'
paḍ.peḍ	'sound of fruits falling lightly down one after another', or 'chickens dying one after another'
buŋ.buŋ	'sound kicking beetle makes'

(81) Sora

geʔ.geʔ	'cry of wild fowl while soaring in sky'
dirre.rirre	'sound to call a dog'
ladur.ladur	'snoring'
rad.rad	'sound of country cart wheels'
pim.piŋ	'produce tinkling bell sound'
pajud.pajud	'smack of whip'
diŋ.diŋ	'banging or rattling sound'
deb.deb	'sound of drumming'

A subset of this probably can be considered to be the number of names of birds with total reduplication in both Ho and Sora.

(82)	Ho	
	chep.chep	'bird sp.'
(83)	Sora	
	ŋe.ŋe=kur=tid-ən	'bird sp.'
	ted.ted=la:=tid-ən	'snipe'

5.2.3 Ideophonic (ad)verbs

Particularly characteristic of Munda expressive discourse is the use of reduplicated forms of ideophones in (ad)verb[i]al function, encoding such semantic nuances as the motion of overly small or large things. Many forms with this type of reduplication may be used as both a predicate or as a modifier of a predicate or arguments. The distinction between adverb and verb with respect to lexemes in most Munda languages is not really relevant, only their syntactic functions. Exceptions include forms that are overtly marked as adverb/non-finite, as in the second and third examples in (84) which are marked by the non-finite adverbial markers *=ga:mle* and *=loge*, respectively.

(84)	Sora	
	so:r.sa:r	'rapid movements of rats and birds'
	lemer.lemer=ga:mle	'manner in which lice move to cause itching'
	le:m.la:m=loge	'manner in which lice move to cause itching'
(85)	Ho	
	tapa.tupu	'waddle along on short legs' (a duck, fat kid)
	seɽeʔ.buɽuʔ	'quick motion and noise of rats nibbling grain'
	radaɖ.rodoɖ	'sound of small creature moving along leaves'
	ɟapa.ɟupu	'walk in slow waddling motion' (a duck)

Related to this are forms that convey disorderly, difficult or rapid movement:

(86)	Sora	
	ted.ted ~ tet.ted	'move swinging sidewise frequently'
	sid.sid.sidlamge	'commotion, tumult, turmoil'
	si:gu.sa:guge	'in a disorderly manner'
	ser.serdəm	'wriggle self through narrow passage'
	sed.sed	'throw into disorder'
	rum.ram	'tread, trot, prance on horse'

rum.rum	'rapid movement in dancing'
obseŋ.oblai	'helter-skelter, in disorder'
obuŋle.abuŋle	'turning side to side'
obuŋten.abuŋten	'rolls lazily'

(87) Ho

tirub.mirub	'go along with head bent down'
ʈaʈub.maʈub	'walk along unsteadily due to weakness at knees'
tara:.turu:	'walk unsteadily because can't see (dark/blind)'
tarab.turub	'move along with head bent down looking down'
solo.bolo	'mix together (people of different castes)'
sigil.bigil ~ sigli.bigli ~ silgi.bilgi	'scatter in disorder'
saul.baul	'become disturbed, and move around in excited manner'
par.pir	'scatter in all directions'
pasa.poso	'quickly disintegrate when touched'
ombo:.sombo:	'slow stride of a fat man'
gaelaŋ.guiluŋ	'move around slowly, mope'
hor.bor	'be in a hurry'
hoɽo.boɽo	'be in a hurry'

Reduplicated expressive forms in Munda languages may also be used to convey various means of looking at someone or something abnormally.

(88) Sora

taŋ.taŋ=mad	'look eagerly at', 'glance with evil eye'

(89) Ho

laɖ.luɖ	'look here and there with big eyes'
lanɖa.lunɖu	'stare out with big round open eyes'
ga:r.gu:r	'gape at here and there' (baby)
guɽu.guɽu	'stare at someone for long time'

Another use of reduplicated ideophonic (ad)verbials is to encode various nuanced means of, defective, problematic or atypical speaking.

(90) Sora

sir.sa:r=tam	'eject small drops of saliva while speaking'
saŋe:.saŋe:	'gasping, panting'
pose:ge.pəse:ge	'in whispers'
ɲerud.ɲerud	'grumble'
ɲid.ɲidber	'stammer, babble'
ɲodi.ɲodi	'chatter'

ɟid.ɟidber-ən	'incoherent talk'
güi.güi	'to murmur'; 'to complain'; 'find fault'
bəduŋ.bədeŋ=tam-ən	'hoarse'; 'in a loud voice'

Note that in Ho, this type of expressive meaning prefers to be encoded by complex reduplication, but not Sora, where total reduplication is also found in such meanings.

(91) Ho

lambaɖ.lumbuɖ	'talk a lot and rapidly'
laɽor.baɽor	'speak indistinctly for a long time, as when drunk or with tongue deformity'
nurum.durum	'talking inaudibly to a third party'
ɟe:ʔr.pe:ʔr	'continuously speak indistinctly with mouth partly open'

Another common meaning of reduplicated expressive ideophonic adverbial forms in Munda languages is to express a range of negative mental states, pain, emotions, etc.

(92) Ho

aka.baka	'be dumbfounded, be confused'
tar:r.tu:r	'utterly confused', 'concussed'
gama:.gumu:	'be shy, feel shame'
ka:la.ko:lo	'completely exhausted from hunger'

(93) Sora

sətid.səboi	'smarting sensation'
gelo.belo	'feel shy and confused'

Some curious, but Munda-characteristic patterns are also found, such as a series of words referring to water- or beer-related terms. Locally made alcoholic beverages are vital to the identity of many Munda-speaking tribal people and a frequent topic of conversation, as are cooking and discussions relating to food in general.

(94) Sora

lakka:da:b.lakka:da:b	'bubbling of boiling water'
lako:dəm.lakodəm	'bubbling sound'
lal.la:r-ən	'scum on surface of liquids'
laduŋ.laduŋ	'sound of tubers boiling in water'

(95) Ho

ʈipa.ʈopo	'enough liquid to be pasty, undrinkable'
ʈip.ʈop	'more watery, drinkable' (rice beer)

saʈa.soʈo	‘sound of rice-beer fermenting’
saɽa.soɽo	‘sound of water rolling over rough land, stones’

Another area of specific concern to Munda speakers addresses oddities in types of soil conditions. Their expressive vocabulary reflects this.

(96) Sora

laku.lake	‘fertile’

(97) Ho

ʈaŋ.ʈoŋ	‘make several breaks in embankment’
sagor.bagor	‘sound of ploughing in a soaked but not flooded field’
baɖa.buɖu	‘lower part of plough to enter soil in soft earth’
bukuɖ.bukuɖ	‘cut easily through soft soil (of plough)’
etom.etom	‘repeatedly leave space between furrows when ploughing’
ɟaɽa.ɟuɽu	‘be thick with moisture so clogs and ploughing is difficult (soil)’

A subset of this type of word refers to the conditions of slopes, paths, or jungle.

(98) Sora

lɨŋ.lɨŋ(dəm)	‘narrow, difficult of a path’
le:ŋ.le:ŋga:mle	‘with a gentle slope’
duŋ.duŋdəm	‘sloping, slanting’

(99) Ho

kuʈi.muʈi	‘elevated land with regular ups and downs’
mela.mili	‘long vista of level treeless land’
uʈu.puʈu	‘covered with dense undergrowth, difficult to pass’ (path, jungle)
rabaɽa.rubuɽu	‘be uneven (terrain)’
hara.huru	‘terrain with many ravines’

Also related to this is the use of expressive reduplication in terms referring atmospheric conditions. Since the sample size is small, it is yet not clear if the use of total reduplication is an artifact of sample size or reflects a real trend that disfavors complex reduplication in this particular semantic domain.

(100) Sora

mom.mo:ri ~ mor.mo:ri:	‘foggy’
mede:r.mede:r	‘dim, dusking’
lil.liŋboŋ-ən ~ liŋ.liŋboŋ-ən	‘rainbow’
jeŋ.jeŋ	‘blow gently, like breeze’

(101) Ho

nili?.nili?	'intense heat'
beɽel.beɽel	'(rain) fall steadily in light drops'
cha?.cha?	'sound of heavy rain'

Another common semantic domain of reduplicated expressive forms refer to properties of light from a non-ideal or non-canonical perspective:

(102) Sora

piɟul.piɟul	'in a sparkling manner'
ɲidur.ɲadurge	'glittering (like cat's of eyes at night)'
ɲidur.ɲudurge	'glittering (like ghost's eyes)'
ɲãdur.ɲãdur	'dazzling'
malaŋ.malaŋ	'to flicker as a flame'
mali:.mala:i	'to be bright, to shine'
da:kul.da:kul	'dimness of view or sight'

(103) Ho

ɟilob.ɟolob	'flashing, sparkling'
ɟilib.ɟilib ~ ɟilib.ɟolob	'repeated flashes of lightning'
ɟili.mili	'to sparkle in many colors'
ɟolo.molo	'clear and sparkling'

Note that as the second example in (103) shows, there is typically no functional difference between forms with total reduplication and complex reduplication in the Munda languages. Conversely, if both total reduplication and partial reduplication are found in the same Munda language, there frequently are functional/semantic differences between those two reduplicative strategies. Since all three strategies appear to be old, more research is needed to determine not only what formal strategies for reduplication were available but also what functional oppositions or subsystems these may have entailed in the proto-language.

5.3 Grammaticalized functions (constructions [> affixes])

Both total reduplication and partial reduplication have been grammaticalized or morphologized in various Munda languages at different historical stages. Note that similar formations have been reported in other Austroasiatic groups as well, e.g., by Radhakrishnan (1981) for Muot/Nancowry and Srichampa (2002) for Vietnamese, so may be very old in the phylum.

5.3.1 Aktionsart including pluractional

One of the most common functions of reduplication is to encode *Aktionsart* categories such as iterative, continuative, progressive, and durative. In Remo (105), a lexical verb that is monosyllabic takes a partial reduplication form (CV-) with the auxiliary verb *ḍen* to mark imperfective categories like the progressive (104).

(104) Lexical verb<REDUPLICATED NON-FINITE FORM> Auxiliary Verb<FULLY INFLECTED FINITE>

(105) Remo
Niŋ *nsuṛaʔ* *susum* *ḍen-t-iŋ*
I banana REDPL~eat AUX-NPST-1
'I am eating a banana' [Field Notes]

The auxiliary verb *ḍiŋ* also requires partial reduplication of monosyllabic lexical verbs (C-) in Hill Gtaʔ.

(106) Hill Gtaʔ
tʃtʃoŋ n-ḍiŋ-we
REDPL:eat-1-AUX-FUT
'I am eating' [Field Notes]

In Juang, the progressive suffix or clitic ***-nɔm*** requires monosyllabic stems to take an allomorph with total reduplication.

(107) REDPL-Verb-TAM$_I$-TAM$_J$(-SUBJ) < *REDPL-Lexical Verb+Auxiliary Verb-TAM$_J$-(SUBJ)

(108) Juang
aiɲ ***dʒɛ'gdʒɛ'g-nɔm****-an*
I REDPL.cry-PROG-PST.ITR
'I was weeping' [Pinnow 1960-ms: 122]

Sora ***lo:-*** and Kharia ***lo*** are likely cognate and in both languages they are formally clitics and not affixes. The element, which encodes frequentative in Sora but continuous in Kharia, obligatorily requires monosyllabic stems to take an allomorph with total reduplication in both. This suggests such a construction in very old in Munda.

(109) Sora
gu-gu-lo:*-te-n*
REDPL-call-FREQ-NPST-ITR
'he calls (me) frequently' [Ramamurti 1931: 28]

(110) Kharia
***gamgam-lo**ʔ-ki-maj*
REDPL:talk-CONT-PST.I-PL
'they kept on talking' [Malhotra 1982: 145]

Pluractional marked verb forms (Cabredo-Hofherr & Laca 2012, Cable 2012; Wood 2007, Xrakovskij 1997) typically have verbs with partial C_1-reduplication in Gtaʔ (1, 111).

(111) Plains Gtaʔ [Anderson in preparation-b]

ho-baʔlir-tʃe-ka	*bari*	*ttar-tʃe*	*ḍḍo-ḍiŋ-har-ge*
RCP-talk-SS-DISC	DISC	REDPL~come.out-SS	REDPL~run-IPFV-PLURACT-PST

'after they discussed this, they came out and (started) running away'

Pluractional, distributed, repeated, iterative, and continuous action can be expressed lexically in forms that are inherently reduplicated as well, as in the following forms from Ho (112).

(112) Ho

baṭa.baṭi	'to fall over repeatedly'
biṭil.biṭil	'repeatedly move body up and down'
chẽ.chẽ	'sound of one baby crying repeatedly'
chã.chẽ	'sound of many baby crying at once'
daḍ.duḍ	'beat something repeatedly to raise dust'
heker.peker	'repeatedly shake due to old age or fear'
ɟeːʔr.peːʔr	'continuously speak indistinctly with mouth partly open'
ṭoːnḍ.ṭoːnḍ	'drag feet repeatedly, continuously'
ṭoːnḍ.moːnḍ	'drag feet while walking'
dəkuta.dəkuta-n	'stay continuously'

5.3.2 Valence

Both the causative *ab-* (113) and reciprocal *al-* (114) require total reduplication with monosyllabic roots/stems in Sora.

(113) Sora

abɟid.ɟid	'affix, attach gum'
abder.der	'persuade, assure, convince'; 'cheat, delude'
abded.ded	'shake head in disapproval'

Note that with the reciprocal, sometimes the prefix itself is copied (the final two examples below), but under our present state of knowledge, it is not clear when and why this copying occurs and when and why it does not.

(114) Sora

allo:.lo:	'engage laborers to rake clearing with hoes'
alkaɲ.kaɲ(-ən)	'abuse each other' ('altercation')
alrundaŋ.alrundaŋ	'pushing, jostling each other'
alguŋ.alguŋ	'vie, compete with each other'

In Ho, a construction consisting of the reciprocal infix *-p-* plus a reduplicated stem in the form X X/*p*/ is found in a number of lexemes with reciprocal semantics (115).

(115) Ho

*baḓ.**bep**aḓ*	'to insist in asking, giving'
*denga.**dep**enga*	'(mutual) help'
*erel.**ep**erel*	'envy, resentment'
*harom,**hap**arom*	'collection of contributions for village sacrifice'
*taŋa.**tap**aŋa*	'separation'

5.3.3 Distributive numerals, pronominals

Distributive forms of at least lower numerals are formed by reduplication in Ho. Total reduplication is found with *apan* (116) while the numeral stems *mi[ḓ]* 'one' and *bar* 'two' undergo partial CV- reduplication to form distributives (117).

(116) Ho

apan.apan	'each his own'

(117) Ho

ba:.bar	'two each', 'two by two'
mi.mi	'one each'
mi.mi:ja	'very many to each one'

5.4 Sora baby talk

Sora baby talk is characterized by a significantly high use of partial and total reduplication in substitutive forms not found in adult speech – a phenomenon commonly attested in child language (Dressler et al. 2005). However, the study of Sora baby talk remains in its infancy and it is not yet clear how and where or why partial reduplication happens instead of the expected full reduplication.

(118) Sora

pa:pi:-n	'vagina' (baby talk)
parpar	'sacrifice' (baby talk)
ɲumɲum	'to piss' (baby talk)
mimi'd	'sleepy' (baby talk)
jojo:-n	'fish' (baby talk)
kukum-ən	'rat' (baby talk)

6 Summary

Reduplication is an integral part of both the lexicon and grammar of the Munda languages. Formally speaking, reduplication can be total or partial, but the most characteristic feature of reduplication in Munda languages can be considered the various complex or substitutive reduplication patterns with consonant or vowel replacement (or indeed both at the same time). In total reduplication, the reduplicant is an exact copy of the base, and can be one-, two- or three syllables typically. Partial reduplication is typically CV – in Munda, but can be C – alone in Gtaʔ. Complex reduplication with consonant or vowel replacement on the other hand is widespread in the family as a whole, in particular in Sora and the Kherwarian languages. In Sora, *b-* is the default replacement consonant, while *m-* plays this role in Ho. Non-labials are less common but attested in both languages as well as the replacing consonant in the reduplicant in complex reduplication patterns. With respect to vowels, in Sora, most base vowels except *a* (at least *u*, *i*, *o*, and *e*) can be replaced by *a* in reduplicants in complex reduplication which appears in the pattern Base-Reduplicant. In Ho, *a* is also by far the most common pattern in complex substitutive reduplication, replacing base vowels in *u*, *e*, *o* or *i*, and both Reduplicant-Base and Base-Reduplicant order are attested.

Reduplication performs a variety of grammatical functions in the Munda languages. These include the creation of non-finite, participial or infinitive forms. In the case of monosyllabic verb stems, the reduplicated infinitive form is required on the lexical verb when it appears in an auxiliary verb construction. The semantics of such formations involving the reduplicated infinitive in auxiliary verb constructions typically involve concepts such as imperfectivity or atelicity. Such formations can be found in a number of Munda languages including Gtaʔ, Remo, Gutob, Juang, Kharia and Sora. Reduplicated stem allomorphs are also required for iteratives in Kherwarian languages, and are also frequent but not obligatory in pluractional forms in Ho or Gtaʔ. In this latter language, most lexical stems that are inherently reduplicated, i.e., lack a non-reduplicated simplex form, encode pluractionality or similar concepts of iterativity, continuous or durative action, etc., or canonically occur in groups. The original C_1- partial reduplication pattern that formed pluractionals appears to have been lexicalized and replaced with a different pluractional element *-har-* in Gtaʔ (Anderson in preparation-c). Reciprocal formations show a preference for reduplication in Sora, and a range of reduplication plus reciprocal marking constructions can be found in Ho as well. Finally distributive numerals are typically reduplicated in Ho. Grammaticalized reduplication in Munda languages formally can reflect total reduplication, total reduplication plus infixation or prefixation, or partial reduplication. Complex reduplication does not perform grammatical functions in Munda languages as a whole.

A number of modificational and adverbial formations as well as onomatopoetic expressions referring to the ambient acoustic environment, including bird names and the name of a number of trees and other culturally salient concepts, e.g., soil or atmospheric conditions are inherently reduplicated expressions. Formally speaking, these constructions involve total reduplication, total reduplication with infixation and complex reduplication with consonant or vowel replacement. Possibly uniting these varied meanings, and certainly the most salient lexical function of total or complex reduplication in the Munda languages, and also most noteworthy from a typological perspective, is the creation of reduplicated 'expressive' forms. These are largely ideophonic (ad)verb[i]al formations that describe the characteristics of actors or actions, often negative, pejorative, odd, unusual, atypical, defective or otherwise noteworthy features of such participants and predicated events. There is neither a bi-unique relationship between any functional subtype of expressive reduplication, nor with the formal means to encode them, but one typically does not find partial reduplication in these substitutive formations. Both total reduplication and complex reduplication with replacement of consonants or vowels on the other hand are commonly found in expressive formations in Sora and Ho. Lastly, baby talk in Sora is also frequently reduplicated, but unlike expressive or ideophonic adverbials, baby talk forms allow both total and partial reduplication.

Syntactically, reduplicated forms in Munda languages can perform almost any role. They can function as predicates and arguments in addition to adjuncts, adverbs or modifiers. To be sure, both Sora and Ho have extensive systems, and potentially even better documented ones could be extracted from large extant lexical copora for Santali and Mundari. Therefore it is hoped that future research will yield a clearer picture of the system of reduplication that characterized Proto-Munda before too long. This in turn will help place Munda within the broader typology of reduplication that characterizes Austroasiatic languages as a whole, and will help us gain a better understanding of how the varied synchronically attested systems found in both modern Munda languages and their more distant linguistic cousins arose and expanded or contracted as they have.

Abbreviations

1, 2, 3	first, second, third person
ATT	attemptive
AUX	auxiliary
CAUS	causative
CLOC	cislocative
CONT	continuous
COP	copula
DISC	discourse marker
FREQ	frequentative
FUT	future
IPFV	imperfective
ITR	intransitive
MDL	middle
NEG	negative
NOM	nominalizer
NPST	non-past
NSF	noun suffix
OBJ	object[ive]
PL	plural
PLURACT	pluractional
PROG	progressive
PST	past
PURP	purposive
RCP	reciprocal
REDPL	reduplication
SS	same subject
SUBJ	subject
TAM	tense-mood-aspect
UND	undergoer

References

Abbi, Anvita. 1992. *Reduplication in South Asian languages: An areal, typological and historical study*. New Delhi: Allied Publishers.

Aboh, Enoch O., Norval Smith & Anne Zribi-Hertz. 2012. Reduplication beyond the word level: A cross-linguistic view. In Enoch O. Aboh, Norval Smith & Anne Zribi-Hertz (eds.), *The morphosyntax of reiteration in creole and non-creole languages*, 1–26. Amsterdam: Benjamins.

Ahmad, Iftikar. 1963. Grammatical reduplication in standard Punjabi. *Pakistani Linguistics*. 65–76.

Alderete, John, Jill Beckman, Laura Benua, Amalia Gnanadesikan, John McCarthy & Suzanne Urbanczyk. 1999. Reduplication with fixed segmentism. *Linguistic Inquiry* 30(3). 327–364.

Anderson, Gregory D. S. in preparation-a. Dictionary of Gta. Plains/Riverside Gtaʔ and Hill Gtaʔ.

Anderson, Gregory D. S. in preparation-b. Gtaʔ Texts: Hill Gtaʔ and Plains Gtaʔ dialects.

Anderson, Gregory D. S. in preparation-c. A grammar of Gtaʔ.

Anderson, Gregory D. S. in preparation-d. Gutob and Remo: Two Munda languages. Grammar, texts, lexicon, history.

Anderson, Gregory. D. S. & K. David Harrison. 2008. Sora. In Gregory D. S. Anderson (ed.), *The Munda languages*, 299–380. Abingdon/Oxford: Taylor & Franicis.

Apte, Mahadeo L. 1968. *Reduplication, echo formation and onomatopoeia in Marathi*. Poona: Deccan College Post-Graduate Research Institute.

Ashirvadam, K. 1992. The structure of Didayi, an Austroasiatic language. Osmania University, Ph. D. Dissertation.

Bakker, Peter & Mikael Parkvall. 2005. Reduplication in pidgins and creoles. In Bernhard Hurch (ed.), *Studies on reduplication*, 511–531. Berlin/New York: Mouton de Gruyter.

Benjamin, Geoffrey. 1976. An outline of Temiar grammar. In Phillip Jenner, Laurence C. Thompson & Stanley Starosta (eds.), *Austroasiatic Studies, Part I*, 129–187. Honolulu: University of Hawai'i Press.

Bhaskararao, Peri. 1977. *Reduplication and onomatopoeia in Telugu*. Pune: Deccan College.

Cabredo-Hofherr, Patricia & Brenda Laca (eds.) 2012. Introduction – Event plurality, verbal plurality and distributivity. In Patricia Cabredo-Hofherr & Brenda Laca (eds.), *Verbal plurality and distributivity*, 1–24. Berlin: De Gruyter.

Cable, Seth. 2012. Distributivity, pluractionality and reciprocity across languages. http://people.umass.edu/scable/LING753-SP12/. (accessed July 17, 2015).

Deeney, John. 1978. *Ho-English dictionary*. Ranchi: St. Xavier Publications.

Diffloth, Gérard. 1976. Expressives in Semai. In Phillip Jenner, Laurence C. Thompson & Stanley Starosta (eds.), *Austroasiatic Studies, Part I*, 249–264. Honoloulu: University of Hawai'i Press.

Diffloth, Gérard. 1979. Expressive phonology and prosaic phonology in Mon-Khmer. *Studies in Tai and Mon-Khmer phonetics and phonology in honour of Eugenie j. A. Henderson*, 49–59. Bangkok: Chulalongkorn University Press.

Downing, Laura. 1998. Onset motivated overcopy in reduplication. In Elly van Gelderen & Vida Samiian (eds.), *Proceedings of the Xth Western Conference on Linguistics*, 81–96. Fresno, California: California State University Press.

Dressler, Wolfgang U., Katarzyna Dziubalska-Kołaczyk, Natalia Gagarina & Marianne Kilani-Schoch. 2005. Reduplication in child language. In Bernhard Hurch (ed.), *Studies on reduplication*, 455–474. Berlin/New York: Mouton de Gruyter.
Emeneau, Murray B. 1938. An echo-word motif in Dravidian folk tales. *Journal of the American Oriental Society* 58. 553–570.
Emeneau, Murray B. 1969. Onomatopoeia in the Indian linguistic area. *Language* 45(2). 274–299.
Gomango, Opino. 2015. Noun morphology in Sora. MA thesis, Deccan College, Pune.
Hurch, Bernhard (ed.). 2005. *Studies on reduplication*. Berlin/New York: Mouton de Gruyter.
Hurch, Bernhard & Veronika Mattes. 2007. The Graz database on reduplication. *Faits des langues* 29. 191–202.
Hurch, Bernhard & Veronika Mattes. 2009. Typology of reduplication: The Graz database. In Simon Musgrave, Martin Everaert & Alexis Dimitriades (eds.), *The use of databases in cross-linguistic studies*, 301–328. Berlin/New York: Mouton de Gruyter.
Inkelas, Sharon & Cheryl Zoll. 2005. *Reduplication. Doubling in morphology*. Cambridge: Cambridge University Press.
Mahapatra, Khageshwar with Dobek Pujari and P. K. Panda. 1989. *Didayi* (In Oriya). Academy of Tribal Dialects and Culture, Government of Orissa, Bhubaneshwar.
Malhotra, Veena. 1982. The structure of Kharia: A study in linguistic typology and change. JNU PhD Dissertation.
Marantz, Alec & Caroline Wiltshire. 2000. Reduplication. In Geert Booij, Christian Lehmann & Joachim Mugdan (eds.), *Morphologie–Morphology*, 557–566. Berlin: Walter de Gruyter.
Moravcsik, Edith A. 1978. Reduplicative constructions. In Joseph Greenberg (ed.), *Universals of human language*, vol. 3, 297–334. Palo Alto: Stanford University Press.
Mustafa, Khateeb S. 1979. Echo-word formation in Dakhini. *Indian Linguistics* 40. 236–242.
Nagaraja, K. S. 1984. Reduplication in Khasi. *Indo-Iranian Journal* 27. 189–200.
Parimalangantham, A. 2009. A study of structural reduplication in Tamil and Telugu. PhD dissertation, PS Telugu University.
Pinnow, Heinz-Jürgen. 1960-ms. Grammatik der Juang-Sprache. Unpublished manuscript.
Rabel, Lily. 1976. Sound symbolism and Khasi adverbs. In Harry L. Shorto (ed.), *Southeast Asian Linguistics Studies* 2, 253–262. Pacific Linguistics C42. Canberra: Australian National University.
Radhakrishnan, Rajagopalan. 1981. *The Nancowry word. Phonology, affixal morphology and roots of a Nicobarese language*. Carbondale, Illinois and Edmonton, Alberta: Linguistic Research.
Raimy, Eric. 2000. *The phonology and morphology of reduplication*. Berlin/New York: Mouton de Gruyter.
Ramamurti, Rao Sahib G. V. 1931. *A Manual of the So:ra: (or Savara) language*. Madras: Govt. Press.
Ramamurti, Rao Sahib G. V. 1938. *Sora-English dictionary*. Madras. Govt. Printing Office.
Rozhanskij, Fedor I. 2011. *Reduplikatsija: Opyt tipologicheskogo issledovanija*. Moscow: Rossijskaja Akademija Nauk. Institut Jazykoznanija.
Rubino, Carl. 2005. Reduplication: Form, function and distribution. In Bernhard Hurch (ed.), *Studies on reduplication*, 11–29. Berlin/New York: Mouton de Gruyter.
Schwaiger, Thomas. 2011. The relation between prototypical and marginal morphology: The case of reduplicative constructions. *Linguistica* 51. 121–134.
Selvam, S. 1988. Expressives in Telugu. JNU University M. Phil. Thesis.

Singh, Rajendra. 1982. On some "redundant compounds" in Modern Hindi. *Lingua* 56. 345–351.
Singh, Rajendra. 2005. Reduplication in Modern Hindi and the theory of reduplication. In Bernhard Hurch (ed.), *Studies on reduplication*, 263–281. Berlin/New York: Mouton de Gruyter.
Srichampa, Sophana. 2002. Vietnamese verbal reduplication. In Robert S. Bauer (ed.), *Collected papers on Southeast Asian and Pacific languages*, 448–461. Canberra: Australian National University.
Starosta, Stanley. 1992. Sora combining forms and pseudo-compounding. *Mon-Khmer Studies* 18/19. 78–105.
Stolz, Thomas, Aina Urdze, Julia Nintemann & Marina Tsareva. 2015. When some dots turn a different color ... Thoughts on how (not) to determine whether or not reduplication is universal. *Studies in Language* 39(4). 795–834.
Stolz, Thomas, Cornelia Stroh & Aina Urdze. 2011. *Total reduplication: The areal linguistics of a potential universal*. Berlin: Akademie Verlag.
Wälchli, Bernhard. 2005. *Co-compounds and natural coordination*. Oxford: Oxford University Press.
Wälchli, Bernhard. 2007. Ko-Komposita (mit Vergleich mit Parallelismus und Reduplikation). In Andreas Ammann & Aina Urdze (eds.), *Wiederholung, Parallelismus, Reduplikation: Strategien der multiplen Strukturanwendung*, 81–108. Bochum: Brockmeyer.
Wiltshire, Caroline. 1999. Expressives in Tamil. *Yearbook of South Asian Languages and Linguistics 2*. 119–137.
Wood, Esther J. 2007. The semantic typology of pluractionality. Berkeley: Ph. D. Dissertation, University of California Berkeley.
Xrakovskij, Viktor. 1997. Semantic types of the plurality of situations and their natural classification. In Viktor Xrakovskij (ed.), *Typology of iterative constructions*, 3–64. München: Lincom.

Rita Finkbeiner

Bla(h), bla(h), bla(h). Usage and meaning of a repetitive all-rounder

Abstract: The paper starts from the question whether *bla(h), bla(h), bla(h)* – a linguistic expression that is ubiquitous in everyday conversation – is a case of (non)-prototypical total reduplication. It is argued that this syllable triple fulfills a range of formal criteria of total reduplication. On the meaning side, *bla(h)* is descriptively merely indexical, but may convey a range of additional expressive meaning aspects. However, it seems that triplication of *bla(h)* does not lead to a substantial change in meaning. This is taken as an argument against reduplication. On the other hand, the results of a corpus study in COCA indicate that there is a strong conversational preference to triplicate *bla(h)*. It is concluded that *bla(h), bla(h), bla(h)* is neither an instance of (non)-prototypical reduplication nor of 'free' repetition, but a conventionalized repetitive pattern.

Keywords: conversational implicature, expressive meaning, indexical, pejoration, triplication

1 Introduction

Utterances containing the seemingly content-free syllable triple *bla(h), bla(h), bla(h)* can be heard frequently in everyday conversation. Consider the following examples. (1) is from a news show on American television, (2) from a Swedish newspaper article, and (3) from a disputation between a German teenage girl and her mother.

(1) PETER VAN SANT: But what about Anita's claims that Robert was violent in their marriage? [...] Not so, says Robert's daughter Stephanie, who witnessed the same incident.
STEPHANIE CLINE: She came downstairs and was yelling at me, do you see this? Do you see this? Your dad just slapped me. He hit me. He's crazy. **Blah, blah, blah.** No. That's not what happened at all because I was

Rita Finkbeiner: Fachbereich 05 – Philosophie und Philologie, Deutsches Institut, Deskriptive Sprachwissenschaft, Johannes Gutenberg-Universität Mainz, 55099 Mainz, Germany.
E-mail: finkbein@uni-mainz.de

https://doi.org/10.1515/9783110599329-003

right there. I saw it. He never laid a hand on anybody, he would never do that, dad's not that type of person.

[COCA, Spoken, CBS, The verdict, 2015]

(2) I många år har vi fått höra samma argument från maktens män: "Det är faktiskt kompetens som måste styra **bla bla bla** fler kvinnor borde gå på KTH, **bla bla bla**, det handlar om äganderätt **bla bla bla**". Och hundra andra varianter på dåliga ursäkter.
'For a long time, we have heard the same arguments from men in powerful positions: "It's in fact expertise which is the crucial factor **blah blah blah** more women should go to KTH [The Royal Technical University, R.F.], **blah blah blah**, it's all about ownership **blah blah blah**". And a hundred other variants of bad excuses.' [my translation, R.F.]

[http://www.svd.se/naringslivets-kvinnor-ofta-dubbelt-sa-bra-som-mannen]

(3) Tochter: [ich hab dich ja] schon mal drauf angesprochen ob ich hier hochziehen darf (.) [...] un da hast du ja immer nur so (.) gründe abgegeben so von wegen wenn du achtzehn bist oder (.) umzug so schwer un **bla bla bla** un so (.) ich find du hast irgendwie so blöde gründe gegeben (.) ähm (.) so * wieso ich nich darf und so!
'Daughter*:* [In fact, I have] already talked to you about moving up here [to this other room] [...] and then you always gave these (.) reasons you know like when you are eighteen or (.) hard to move and **blah, blah, blah** and the like (.) I think you somehow gave these silly reasons (.) ah (.) like (.) why I'm not allowed to and things like that!' [my translation, R.F.]

[Datenbank gesprochenes Deutsch, Korpus Elizierte Konfliktgespräche]

The examples indicate that *bla(h), bla(h), bla(h)* is ubiquitous in many languages. We find it not only in English, Swedish, and German, cf. (1)–(3),[1] but also in French, Italian, Spanish, Polish, and Hebrew, and the list may certainly be continued. Interestingly, *bla(h), bla(h), bla(h)* exhibits a similar pronunciation and graphematics, as well as a similar range of meanings in genealogically different languages. What speakers may convey by using *bla(h), bla(h), bla(h)* is not only reference to some previous utterance(s), but also expressive meaning aspects such as irrelevance, 'coolness', and pejoration. As *bla(h), bla(h), bla(h)* is widespread in many languages, we may regard it an international word, or internationalism. However, while most international words (e.g., *hotel, police, microscope*) are loan words that occur in several languages, resulting from sim-

1 Cf. on English *blah* also Persson (1974), on Swedish *bla* Lindström (1999).

ultaneous or successive borrowing from some language of origin (e.g., Latin or Greek), in the case of *bla(h), bla(h), bla(h),* it is difficult to identify its source language. It may have developed in several languages independently, given that it is an onomatopoetic expression, which (universally) imitates the sound of human talk. Both in the sense that *bla(h), bla(h), bla(h)* occurs in many languages, and in the sense that it has several layers of meaning, we may call it a repetitive all-rounder.

However, this is only a first descriptive approximation. In the context of the present volume, the evident question is whether *bla(h), bla(h), bla(h)* is an instance of (grammatical) total reduplication, or whether we are dealing with an instance of 'free' (conversational) repetition. In the following, I will approach this question mainly from a functional perspective, with a focus on American English. In Section 2, I provide a definition of "the canonical type" of total reduplication and try to match our case with this definition. As we will see, *bla(h), bla(h), bla(h)* fulfills a number of formal requirements of total reduplication, but behaves differently as to its meaning. In Section 3, I differentiate between two main usage contexts of *bla(h), bla(h), bla(h),* the monophonic and the polyphonic usage context. In Section 4, I analyze the meaning of *bla(h)* in the two usage contexts, suggesting a network of conceptual paths that connect the different usages to each other. In Section 5, I further investigate the discoursal statuses of the different meaning aspects, drawing on a recent typology of pragmatic interpretations (cf. Ariel 2016). In Section 6, I ask for the role of triplication in the meaning constitution of *bla(h), bla(h), bla(h).* I argue that triplication does not add or change any semantic aspects as compared to the meaning of the singleton item. However, triplication is shown to be a clear conversational preference, which may enhance the expressive aspects of *bla(h), bla(h), bla(h).* Section 7 concludes.

2 A case of total reduplication?

In order to decide whether *bla(h), bla(h), bla(h)* is a case of total reduplication, we may take the recent definition of Stolz & Levkovych (to appear) as our basic definition (cf. also Stolz et al. 2011, Finkbeiner & Freywald to appear).

> The canonical type of reduplication applies if two syntagmatically immediately adjacent linguistic signs, which are identical in form and meaning, form a construction the meaning/function of which is at least slightly different from that of the singleton item which participates in the reduplicative construction. (Stolz & Levkovych to appear, 4)

In their paper, Stolz & Levkovych focus on the delimitation of reduplication from repetition. While both processes may, on the formal side, be characterized by two (or more) adjacent, identical signs, on the meaning side, the main difference is that in reduplication, the meaning of the reduplicative construction is "at least slightly" different from the meaning of the singleton item, while in repetition, there is no proper semantic alteration. That means that reduplication is taken to be a *grammatical* process which is, in a given domain, compulsory. For instance, Breton uses reduplication to form the elative from the positive, cf. *bras* 'big' vs. *bras~bras* 'very big' (Stolz & Levkovych, to appear, 9). This is a grammatical means, not a free choice. By contrast, repetition is taken to be a pragmatic process which is not compulsory, but "a matter of style" (p. 15). For instance, one may repeat the name *Harry* several times – in fact unrestrictedly often – in order to reach a communicative effect, e.g. an effect of particular emphasis.

Now, what about *bla(h), bla(h), bla(h)*? Does this expression fulfill the criteria of total reduplication? On the one hand, this expression is a chain of syntagmatically immediately adjacent, totally identical linguistic signs. This matches part of the above definition on the formal side. On the other hand, the number of occurrences is not restricted to two. What is more, if we look at the meaning side, it is not a trivial task to decide whether there is a difference in meaning between the singleton *bla(h)* and the construction *bla(h), bla(h), bla(h)*. This is in part due to the fact that we have a hard time to say what the (lexical) meaning of *bla(h)* is.

In order to come to grips with the two problems, we first take a closer look at the possibilities regarding the number of occurrences. While triplication, i.e. a chain of three adjacent identical items *bla(h)*, seems to be the prototypical case, a search in the Corpus of Contemporary American English (COCA) reveals that *bla(h)* in fact can be used as a singleton item as well, with approximately the same meaning, cf. (4).

(4) Single occurrence
What's interesting now is we're all nervous about the kids being on the Internet. [...] All you can do is raise a kid who will talk to you if they see something weird. It's like in "Knocked Up" where she says I Googled murder. Well, what did you see? There was some people on the ground and blood and **blah.**
[COCA, Magazine, NPR, 2012, In This Is 40, Family Life In All Its Glory]

Next, we also find double occurrences of *bla(h)*, cf. (5).

(5) Double occurrence
Chris Matthews responded to those passages in the speech and said he made a mistake about cable television, a lot of people at our network es-

pecially, says Chris, who have hoped and shared his hope for ending racial division and **blah, blah.**
[COCA, Spoken, Fox, 2015, Press Tackles the Patriots; Obama Takes on Pundits; Larry Wilmore's Debut]

Finally, there are also multiple occurrences, i.e. chains of more than three adjacent *bla(h)*s in the corpus, cf. (6).

(6) Multiple occurrence
What about Dr. Ben Carson? Somebody needs to check his vital signs. Was he awake? What is his, what is his resting heart rate? He was like **blah, blah, blah, blah, blah, blah.**
[COCA, Spoken, ABC, 2015, Hot Topics]

We may draw the preliminary conclusion that the number of items in the construction under scrutiny is not restricted to two. This is a first indication that *bla(h), bla(h), bla(h)* might not be a prototypical case of reduplication. While the examples show that the number of items is not strictly limited, intuitively, triplicated *bla(h), bla(h), bla(h)* seems to be the prototypical case. We come back to this later.

Let us now take a closer look at the meaning criterion. Is there a difference in meaning between *bla(h)* and *bla(h), bla(h), bla(h)*? In order to decide on this question, we have to specify what triplicated *bla(h), bla(h), bla(h)* means, and what single *bla(h)* means. This is not a trivial question, however. Let's have a look again at our example (1), here repeated as (7) for convenience.

(7) PETER VAN SANT: But what about Anita's claims that Robert was violent in their marriage? [...] Not so, says Robert's daughter Stephanie, who witnessed the same incident.
STEPHANIE CLINE: She came downstairs and was yelling at me, do you see this? Do you see this? Your dad just slapped me. He hit me. He's crazy. **Blah, blah, blah.** [COCA, Spoken, CBS, The verdict, 2015]

What does Stephanie Cline convey with this utterance? There are at least three different aspects of meaning here. First, by using *bla(h), bla(h), bla(h)*, Stephanie points to some utterances previously made by Anita. *Bla(h), bla(h), bla(h)* is thus a dummy element that stands in for some linguistic content uttered in a different context. In a sense, thus, *bla(h), bla(h), bla(h)* means something like "(more) words out there". Second, by using *bla(h), bla(h), bla(h)*, Stephanie conveys that what Anita said in this other context is irrelevant for the current conversational purposes. And third, with this utterance, Stephanie conveys a derogatory stance towards Anita's behavior.

While all those different aspects of meaning are associated with the usage of *bla(h), bla(h), bla(h)* in this context, it seems that the word *bla(h)* has no proper *lexical* content. As used in our examples, *bla(h)* should best be analyzed as member of the category of interjections (cf. Fries 1990): Just as prototypical interjections, e.g. *yuck*, *oh*, *yummy*, or *oops*, *bla(h)* is not inflectable, it is not integrated syntactically into the sentence, and it may represent an autonomous utterance. Interjections normally do not have lexical meanings, but fulfill specific functions in the expression of emotions.

Against this analysis, one might object that in fact, there are both a noun and an adjective *blah* in English that do have lexical content. The noun *(the) blah* (also in the form *(the) blah-blah*$_{N}$) means something like 'silly or meaningless talk, idle talk'. The adjective *blah* means either 'not interesting, dull, boring' or, in combination with the verb *to feel*, 'without energy or enthusiasm' (e.g., *I feel so blah*). From this latter usage, also a plural noun *the blahs* derives, which means 'the feeling of being bored'. However, I would like to argue that both the noun and the adjective have to be kept apart from our interjection. For the nominal *blah*, it is evident that it is semantically based on the interjection ('empty words out there' > 'idle talk'). Thus, the interjection is the primary instance, while the noun is derived from it. As to the adjective, the connection is not so evident. It is most likely that the adjective and the interjection are just homonyms, the adjective going back to French *blasé* (cf. Merriam-Webster.com). For our purposes, thus, the noun *blah* and the adjective *blah* can be neglected. They do not help us in deciding whether there is a semantic difference between the interjection *bla(h)* and its triplicated form *bla(h), bla(h), bla(h)*.

Thus, at this point we cannot decide whether the triplication makes a difference on the side of semantics. However, what we can do is ask whether triplication makes a difference on the side of usage, or pragmatics. Therefore, we will take a closer look in the next Section at the usage potential of *bla(h), bla(h), bla(h)*.

3 Usage contexts

We may distinguish at least the following three usage contexts of *bla(h), bla(h), bla(h)*: The monophonic usage, the polyphonic usage, and the turn-initial reply usage. In this paper, I will focus on the monophonic and the polyphonic usages. For more details on the turn-initial reply usage, cf. Finkbeiner (2016).

In the monophonic usage, *bla(h), bla(h), bla(h)* is used in a piece of text or discourse which is produced from the perspective of one single author, who solely is taken to be responsible for the attitudes conveyed in the text. In the polyphonic usage, *bla(h), bla(h), bla(h)* is used in a piece of text or discourse which involves the utterances, perspectives or attitudes ('voices') of several different speakers (cf. on the concept of 'polyphony' Fabricius-Hansen 2002, Günthner 2002). Polyphonic contexts include direct and indirect quotations and their variants, e.g., staged conversation or so-called 'berichtete Rede' (Fabricius-Hansen 2002).

An example of a monophonic usage of *bla(h), bla(h), bla(h)* is (8).

(8) Whilst editing, I have noticed that in some articles the dates are formatted as follows: ... on_26_April*ampersand*nbsp; 1997_ at... which produces this: on 26 April 1997 at... What is the advantage? Can somebody help me please? Gareth Griffith-Jones (talk) 18:10, 2 March 2012 (UTC)
That's a non-breaking space. It tells the browser to avoid laying out the text like this...
blah blah blah blah blah blah blah blah blah 26 April
1997 **blah blah blah blah blah blah blah blah blah**
[...] John of Reading (talk) 19:10, 2 March 2012
[https://en.wikipedia.org/wiki/HElp_talk:Editing]

In this piece of (written) dialogue from a Wikipedia discussion on text editing, the user "John of Reading", by using *blah, blah, blah, blah, blah ...*, does not quote words or attitudes of other speakers. Rather, he uses *blah, blah, blah, blah, blah ...* just to demonstrate certain typesetting features. In this context, all that matters is the material form of words, not their contents. The empty syllable *blah* is particularly apt for this communicative purpose.

Another example of a monophonic usage is (9).

(9) ROBERT DAVIS: That particular judge, I didn't know that she was the toughest judge they had with all the stuff that we had – police officers had done wrongly to females, she wasn't going look too kindly. You know, there was a lot of press there, news people **blah**, **blah**, **blah**, you know, that kind of stuff. And people were in the hallways looking at me and whispering and so forth. [COCA, Spoken, NPR, 2015, Cop out]

In this example, the police officer Robert Davis, who is accused for sexual abuse of a woman, uses *blah, blah, blah* in his narration in order to indicate that the list of journalists who were at court – *a lot of press, news people, ... that kind of stuff* – could be continued. *Blah, blah, blah* thus functions as a general extender in this example. It is a monophonic usage, because Robert Davis is not referring by this expression to something someone else said in a different context.

An example for a polyphonic usage is our well-known example (1), here repeated as (10) for convenience.

(10) STEPHANIE CLINE: She came downstairs and was *yelling at me*, do you see this? Do you see this? Your dad just slapped me. He hit me. He's crazy. **Blah, blah, blah.** [COCA, Spoken, CBS, The verdict, 2015]

In using *blah, blah, blah,* Stephanie refers to something Anita said in a previous context. Therefore, it is a polyphonic (or: quotational) usage that involves reference to utterances of other speakers. This is linguistically marked by its embedding under a *verbum dicendi* (*she ... was yelling at me ...*).

Another example of a polyphonic usage is (11).

(11) [...] *she seemed to go on about* every other detail anyone could possibly come up with, right down to, and including, the fact that one of the little red lines was pink and that threw the other five positive tests into doubt for them, so they went to the doctor the next day for a blood test, and... **blah blah blah**. The story was endless.
[COCA, Fiction, Harbison, Elizabeth M.: If I could turn back time. New York: St. Martin's Press, Edition: First edition. 2015]

In this example, the protagonist of the novel tells about a friend talking endlessly about how she found out about her pregnancy. By *blah blah blah*, the narrator refers to utterances of this friend, thereby conveying that she found this story – or more generally, this behavior – really annoying.

The examples indicate that there are systematic meaning differences between the two context types: While monophonic usages of *blah, blah, blah* tend to be non-pejorative, polyphonic usages tend to be pejorative. This has to do with the fact that polyphonic contexts often are charged with attitudes and feelings of the different speakers, while monophonic contexts, because of their mono-perspective, are not. On the other hand, monophonic usages often include a speaker-related aspect of 'coolness', i.e. the speaker presents herself as not caring too much about details, in this sense being 'cool' (cf. Wiese & Polat 2016). In the next section, we will go into more detail with these meaning differences, taking a closer look at the expressive aspects of irrelevance, 'coolness', and pejoration.

4 Descriptive and expressive meaning

Let us first try to pin down the meaning of *blah, blah, blah* in the monophonic usage, e.g. (9) (see above). There are three aspects involved in Robert Davis' utterance of *blah, blah, blah*:

(i) 'I could say more/I could be more specific'
(ii) 'The details are irrelevant for the current conversational purposes'
(iii) 'I am cool'

That there is indeed an aspect of signaling 'coolness' becomes evident if we replace *blah, blah, blah* in Robert's utterance by *etcetera*. *Etcetera* also carries the aspects (i) and (ii), but it does not carry the 'coolness' component.

In the polyphonic usage, which involves quotation (e.g., [10]), we also find the aspects (i) and (ii). However, the 'coolness' aspect seems to be toned down here. Instead, there is a prominent aspect of pejoration. Thus, Stephanie Cline conveys the following by uttering *blah, blah, blah* in (10).

(i) 'I could say more/I could be more specific'
(ii) 'The details are irrelevant for the current conversational purposes'
(iii) 'I disdain what Anita said'

In order to systematize these findings, it is useful to adopt the approach of Potts (2007) to expressive meaning. Potts (2007) distinguishes between two kinds of content, namely, descriptive content and expressive content. These two kinds of content can be described independently. For example, an utterance such as (12) can be analyzed as follows:

(12) Nietzsche was a Kraut.
a. Descriptive content: Nietzsche was a German.
b. Expressive content: Nietzsche was {a bad person, in the opinion of the speaker}

According to Potts, the expressive content is not truth-evaluable, as it does not contribute to what is said. That's why the reply in (13B) fails (cf. Potts 2007; Finkbeiner et al. 2016). By uttering (13B), the speaker cannot negate the expressive content of *Kraut*. On the other hand, the expressive content of *Kraut* is not cancellable, as shown in (14). Therefore, it cannot easily be analyzed as a conversational implicature, either.

(13) A: Nietzsche was a Kraut. – B: #No, that's not true.

(14) #Nietzsche was a Kraut, but I don't want to convey that Nietzsche was a bad person.

For the case of *bla(h), bla(h), bla(h)*, we may adopt Potts' distinction as follows. The descriptive content of *bla(h), bla(h), bla(h)* is that it indexically indicates that something has been said or could be said in a speech context other than the current one. The expressive content is threefold. First, the speaker may convey an evaluation of that content as irrelevant. Second, a speaker may present herself as 'cool', and third, a speaker may convey a derogatory attitude towards the omitted content, or towards its originator.

The picture we arrive at is the following:

Layers of meaning in *bla(h), bla(h), bla(h)*
- Descriptive (indexical) component: 'Something has been said/will be said/might have been said/could be said' ("words, words, words out there")
- Expressive component 1: 'The speaker considers the content of what has been said/will be said/might have been said/could be said irrelevant for the current purposes.'
- Expressive component 2: 'The speaker presents herself as 'cool''
- Expressive component 3: 'The speaker/writer disdains what has been said/will be said/might have been said/could be said.'

This is, so far, a plain description of what speakers may convey by using *bla(h), bla(h), bla(h)*. However, we also should make an effort to find out how the different (descriptive and expressive) meaning aspects are related to each other. After all, they are not just a random collection of aspects, but can be regarded part of a systematic network of conceptual relations. In this network of meanings, both the status of *bla(h)* as an onomatopoetic interjection and more general conversational maxims play a crucial role. We may sketch this network of relations as follows, cf. Figure 1 below.

First of all, *bla(h)* is an onomatopoetic interjection that imitates the sound of human talk, indicating that something was said or could be said in a context different from the present one. This gives us the descriptive (indexical) meaning component. In imitating sounds only, the speaker does not make reference to content. This is clearly an instance of marked language use: Normally, we use language to convey contents. If a speaker uses language just to imitate sounds, this is a violation of Grice's Maxim of Manner, or the M-Principle (Levinson 2000). The M-principle says, in short, that a marked message indicates a marked situation. If we take it that the speaker is cooperative, then we can infer the M-implicature that the speaker used marked language (no content, just sound) exactly because she wanted to indicate that the omitted content is not relevant. This gives us the meaning component of irrelevance.

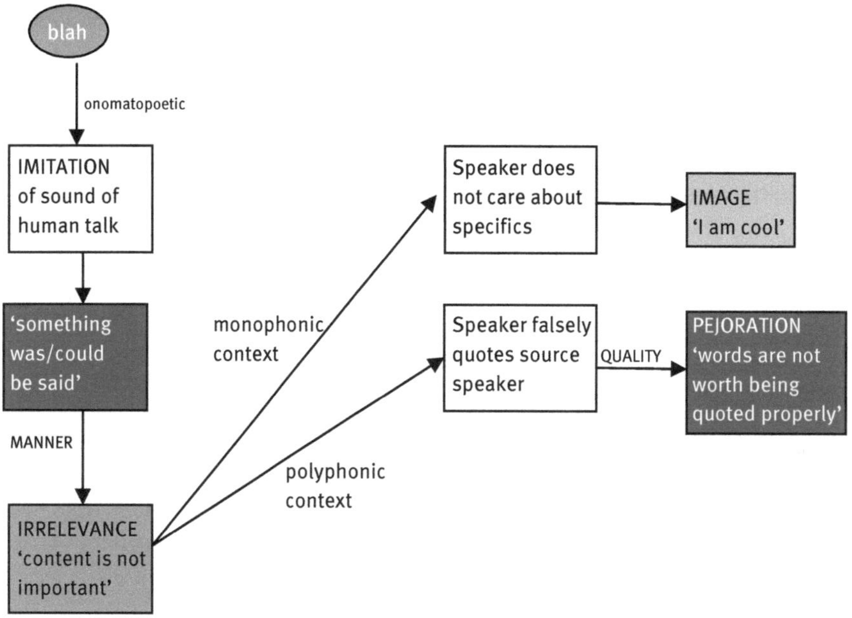

Figure 1: Network of meaning components.

Next, we have seen that speakers can use *bla(h)* in monophonic or in polyphonic contexts. If *bla(h)* is used in a monophonic context, the component of irrelevance must be interpreted as mainly speaker-related. A speaker who is inattentive to detail may be perceived as someone who just does not need to care about specifics, and therefore as someone who is 'cool'. This gives us the expressive component of 'being cool' (cf. Wiese & Polat 2016). By contrast, if *bla(h)* is used in a polyphonic context, its use amounts to a false quotation of some other speaker. Normally, the Maxim of Quality would require that what we say is true. Thus, as quotational speakers, we are expected to render the source speaker's utterance as correctly as possible ('verbatim assumption', Clark & Gerrig 1990). However, in quoting the words of someone else by saying *bla(h)*, a speaker blatantly violates the Maxim of Quality. This may give rise to a pejorative implicature, namely the implicature that the speaker thinks that the words of the source speaker are not worth being quoted properly, e.g., because they were silly, prolix, dull or the like. This gives us the expressive component of pejoration.

5 Discoursal status of the meaning components

So far, we have treated irrelevance, 'coolness', and pejoration as three different expressive meaning aspects that may arise in utterances of *bla(h)*. However, from the point of view of the semantics/pragmatics distinction, the question is whether they all have the same status when it comes to their strength, or discourse-prominence. In the following, I will take a closer look at the discoursal status of these three expressive components, using the test battery suggested by Ariel (2016). The hypotheses are that irrelevance, on the one hand, is a strong meaning component (or an *explicature*) nearly inseparable from the linguistic meaning of *blah, blah, blah*, while 'coolness' and pejoration, on the other hand, are weaker, separate, additional meaning components (or *implicatures*).

Ariel (2016) suggests the following tests (among others) to decide whether a meaning component of an utterance is an explicature or an implicature, cf. Table 1. According to Ariel (2016), an explicature passes the Said test and the That-is test but fails in the Indirect-addition test, whereas an implicature fails on the Said test and the That-is test, but passes the Indirect-addition test. The tests are named after the type of paraphrase one can use to faithfully report on the utterance in question.

Table 1: Tests for the discoursal status of meaning components (cf. Ariel 2016). [✓: pass, X: fail]

Test	Explicature	Implicature
Said	✓	X
That-is	✓	X
Indirect-addition	X	✓

Let us start with the meaning component of irrelevance ('the details are irrelevant for the current purposes'). We take the Robert Davis example, again, repeated as (15) for convenience.

(15) ROBERT DAVIS: You know, there was a lot of press there, news people **blah, blah, blah**, you know, that kind of stuff. And people were in the hallways looking at me and whispering and so forth.

The irrelevance component passes the Said test:

Said test:

✓ Robert **said that** there was a lot of press there, news people **and others whom to detail is irrelevant for the current purposes. ✓**

We may faithfully report on what Robert Davis said by saying *Robert said that* ..., thereby replacing *blah, blah, blah* by a paraphrase of the irrelevance component. The irrelevance component also passes the That-is test, in which we keep *blah, blah, blah,* but add a *that is* ... sentence.

That-is test:

✓ Robert **said that** there was a lot of press there, news people *blah, blah, blah,* **that is, news people and others whom to detail is irrelevant for the current purposes.**

However, it is weird to rephrase the example such that we report on what Robert said by a sentence that states that the irrelevance aspect was an *indirect addition* to what was said:

Indirect-addition test:

X Robert **said that** there was a lot of press there, news people blah, blah, blah. **In addition, Robert indirectly conveyed that there were also others whom to detail is irrelevant for the current purposes.**

This does not seem to be a faithful report. Next, we take the 'coolness' component ('I am cool'). This component clearly fails on the Said test.

Said test:

X Robert **said that** there was a lot of press there, news people **and (he said) that he is/was cool.**

It also fails on the That-is test.

That-is test:

X Robert **said that** there was a lot of press there, news people *blah, blah, blah,* **that is, (Robert said) that he is/was cool.**

However, the 'coolness' component seems to pass the Indirect-addition test.

Indirect-addition test:

✓ Robert **said that** there was a lot of press there, news people blah, blah, blah. **In addition, Robert indirectly conveyed that he is cool.**

Finally, we check the pejoration component ('I disdain what has been said'). For this component, we have to use an example that contains a pejorative component, cf. our Stephanie Cline example, repeated as (16).

(16) STEPHANIE CLINE: She came downstairs and was yelling at me, do you see this? Do you see this? Your dad just slapped me. He hit me. He's crazy. **Blah, blah, blah.**

The pejorative component fails on the Said test.

Said test:

X Stephanie said that Anita came downstairs and was yelling at her, do you see this? Do you see this? Your dad just slapped me. He hit me. He's crazy. **And (Stephanie said) that she disdains/disdained what Anita said.**

It also fails on the That-is test.

That-is test:

X Stephanie said that she came downstairs and was yelling at her, do you see this? Do you see this? Your dad just slapped me. He hit me. He's crazy. Blah, blah, blah. **That is, Stephanie said that she disdains/disdained what Anita said.**

However, the pejoration component passes the Indirect-addition test.

Indirect-addition test:

✓ Stephanie said that Anita came downstairs and was yelling at her, do you see this? Do you see this? Your dad just slapped me. He hit me. He's crazy. Blah, blah, blah. **In addition, Stephanie indirectly conveyed that she disdains/disdained what Anita said.**

Thus, we may conclude from our tests that the irrelevance component is an explicature, i.e. a strong meaning component nearly inseparable from the linguistic meaning of *blah, blah, blah*. By contrast, 'coolness' and pejoration have the status of implicatures, i.e. they are weaker, separate, or additional meaning components. While in our examples, we found mostly either 'coolness' or pejoration – depending on monophonic or polyphonic usage context –, this is not to say that 'coolness' and pejoration strictly exclude each other. The borderline between these meaning aspects – being implicatures – is not sharp, and the different aspects may overlap.

We have now gained a more detailed picture of the pragmatic meaning of *bla(h), bla(h), bla(h)* as conveyed in monophonic and polyphonic usages. However, we still do not have an answer to the question whether *bla(h), bla(h), bla(h)* is an instance of (non)-prototypical reduplication. Therefore, we take another look at the role of triplication in the next section. The basic question is whether the aspects of irrelevance, 'coolness', and pejoration are induced by the triplication or whether they occur independently of triplication.

6 The role of triplication

If the expressive meaning aspects occur only with duplicated or triplicated *bla(h)*, this may be taken as an argument in favor of a reduplication analysis, because this would indicate a change in meaning compared to the single item. However, if we find the respective meaning aspects also with single-item *bla(h)*, this may be taken as an argument in favor of a repetition analysis, because we then would not have a change in meaning.

Taking another look at the COCA examples, it seems that triplicating *bla(h)* in a way is superfluous. First of all, there is no meaning difference between duplicated and triplicated *bla(h)*. For instance, we could replace the duplicated *blah, blah* in (17) by triplicated *blah, blah, blah* without causing any meaning difference.

(17) I'm on the fifth floor and the heat is controlled in the nether regions and the radiators never cool, and when I tried to open the bedroom window, it was painted shut. That's when the substitute plumber buzzes. I buzz him in and as he's making his way up I start muttering: goddamn distractions, fuckm heat, got to get some work done, only the work matters, **blah, blah,** and by the time I hear the elevator chunk open it's all his fault, I slide the four deadbolts and swing open the heavy industrial door, a bear of a man not tall, I avoid eye contact, tell him, in the kitchen, first left. [COCA, Fiction, Petronio, B.: The substitute plumber. 2004]

Second, a look at utterances with single-item *blah* reveals that irrelevance, 'coolness', and pejoration may be conveyed even by these utterances. For example, the *blah* utterance in (18) conveys both irrelevance and 'coolness', and the *blah* utterance in (19) conveys both irrelevance and pejoration.

(18) What's interesting now is we're all nervous about the kids being on the Internet. [...] All you can do is raise a kid who will talk to you if they see something weird. It's like in "Knocked Up" where she says I Googled murder. Well, what did you see? There was some people on the ground and blood and **blah.**
[COCA, Magazine, NPR, 2012, In This Is 40, Family Life In All Its Glory]

(19) Got the letter today and guess what: still not a superhero. Dear Applicant, not a good sign, the number of qualified candidates this year blah blah far exceeded the number of available **blah**. I scan the list of people who did make it. A lot of them graduated with me. It's the usual assortment of the strong and beautiful. About half are fireball shooters. A few

> are ice makers. Half a dozen telepath/empaths. A couple of brutes, a shape-shifter, a few big brains. One thing they all have in common is that every single one of them can fly. I can't fly. (Third class superhero)
> [COCA, Fiction, Yu, Charles: Third class superhero. Orlando: Harcourt, 1st ed. 2006]

Thus, we may conclude that triplication of *blah* is not compulsory, but an option speakers have. Clearly, triplication contributes an iconic aspect of 'more of the same'. This may be very useful in contexts in which speakers want to stress that there is 'a lot of content' one could (but for some reason does not want to) specify.

How do these observations fit into our conceptual network of meaning components? We may sketch the role of triplication within this network as follows, cf. Figure 2.

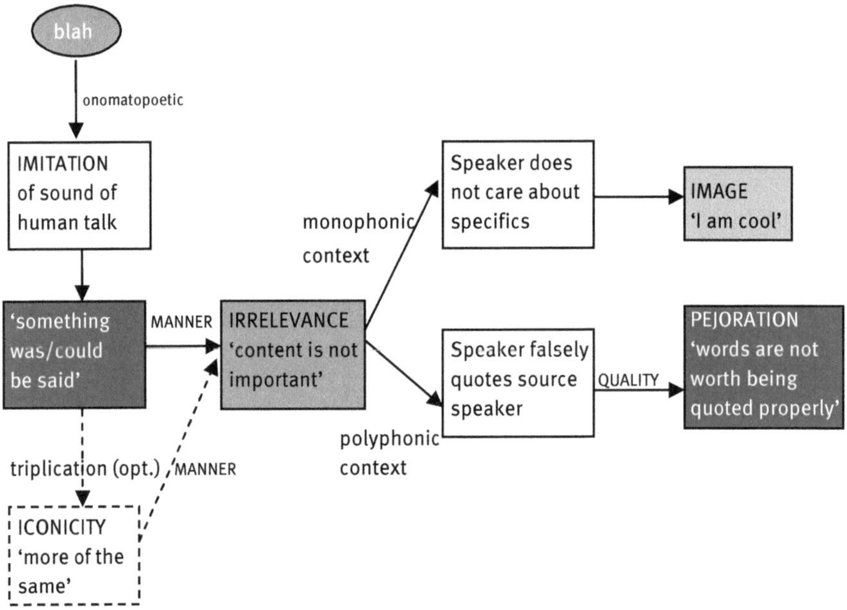

Figure 2: Modified network of conceptual relations.

In Figure 2, triplication is conceived of as a 'loop way', that is, an option by which speakers may enhance the impression of 'more of the same'. However, triplicating *bla(h)* does not add anything on the level of descriptive or expressive meaning.

Still, it would be mistaken to say that triplication is just a random choice speakers may or may not opt for. From Table 2, it becomes evident that triplicated *blah* is by far the most frequent manifestation in discourse, both spoken and written, while duplicated and singleton *blah* occur very rarely.[2] This means that triplication cannot be a free choice, or a mere "matter of style". Thus, it is not a clear-cut case of repetition. Rather, triplication is a strong conversational preference, and as such highly conventionalized. What we are dealing with, then, is neither prototypical reduplication nor prototypical repetition, but a highly conventionalized discourse pattern.

Table 2: Distribution of triplicated, duplicated, and single-item *blah*, COCA Corpus 1990-2015 (520 million words).[3]

	Spoken	Newspaper
Blah, blah, blah	212	47
Blah, blah	8	4
blah	5	4

It is an interesting question why speakers prefer the number three of all possible numbers of occurrences. This is true not only for the case of *bla(h), bla(h), bla(h)*, but also for many other word triples. In German, alongside *bla, bla, bla*, one can find, e.g., the word triples *und, und, und* ('and, and, and'); *na, na, na* ('ts, ts, ts'); *ha, ha, ha* ('ha, ha, ha') and *toi, toi, toi* ('knock on wood!'). I can only touch upon an answer here. Apart from number symbolic, this preference may have to do with the general rhetoric effects of mentioning three of a kind. Two of a kind make two, but three make a list – or a rule, a pattern, or a system. And therefore, more than three would be uneconomic. One can find this 'rhetoric of three' in a great many domains, from three dots signaling ellipsis in graphematics, over three bullet points on lists, three examples in argumentative texts, three sections in a scientific paper, until three episodes in jokes or fairy tales. Thus, triplication is a powerful rhetorical strategy, enhancing conventionalization of triples.

2 In addition, the table reveals that *blah, blah, blah* in fact is much more common in spoken discourse than in written language. This confirms the intuition that it is primarily a *conversational* phenomenon.

3 Note that the numbers include spellings without comma.

7 Conclusion

I have argued in this paper that *bla(h), bla(h), bla(h)* is a linguistic expression that is lexically empty, but has indexical meaning as well as a range of expressive meaning aspects. However, the same meaning aspects may also be conveyed by utterances of singleton *bla(h)*. Therefore, *bla(h), bla(h), bla(h)* is not a prototypical instance of reduplication. At the same time, we have seen that the triplication is not a totally free choice, but a conversational preference. Therefore, *bla(h), bla(h), bla(h)* cannot be regarded a prototypical instance of repetition. I have suggested here to analyze *blah, blah, blah* as a conventionalized repetitive pattern, i.e. a pattern that has its origin in conversation, but is undergoing a process of conventionalization, which may lead, ultimately, to its becoming part of grammar.

References

Ariel, Mira. 2016. Revisiting the typology of pragmatic interpretations. *Intercultural Pragmatics* 13(1). 1–35.

Clark, Herbert H. & Richard J. Gerrig. 1990. Quotations as demonstrations. *Language* 66(4). 764–805.

Fabricius-Hansen, Cathrine. 2002. Nicht-direktes Referat im Deutschen. Typologie und Abgrenzungsprobleme. In Cathrine Fabricius-Hansen, Oddleif Leirbukt & Ole Letnes (eds.), *Modus, Modalverben, Modalpartikeln*, 7–29. Trier: Wissenschaftlicher Verlag.

Finkbeiner, Rita. 2016. *Bla, bla, bla* in German. A pejorative construction? In Rita Finkbeiner, Jörg Meibauer & Heike Wiese (eds.), *Pejoration*, 269–299. Amsterdam: Benjamins.

Finkbeiner, Rita & Ulrike Freywald (eds.). to appear. *Exact repetition in grammar and discourse.* Berlin/Boston: de Gruyter.

Finkbeiner, Rita, Jörg Meibauer & Heike Wiese (eds.). 2016. *Pejoration*. Amsterdam: Benjamins.

Fries, Norbert. 1990. Interjektionen und Interjektionsphrasen. *Sprache und Pragmatik, Arbeitsberichte Nr. 17*. 1–43.

Günthner, Susanne. 2002. Stimmenvielfalt im Diskurs. Formen der Stilisierung und Ästhetisierung in der Redewiedergabe. *Gesprächsforschung – Online-Zeitschrift zur verbalen Interaktion* 3. 59–80.

Levinson, Stephen. 2000. *Presumptive meanings. The theory of generalized conversational implicature*. Cambridge, MA: The MIT Press.

Lindström, Jan. 1999. *Vackert, vackert! Syntaktisk reduplikation i svenskan*. Helsinki: Svenska litteratursällskapet i Finland.

Persson, Gunnar. 1974. *Repetition in English. Part I: Sequential repetition*. Uppsala: Acta Universitatis Upsaliensis.

Potts, Christopher. 2007. The expressive dimension. *Theoretical Linguistics* 33(2). 165–198.

Stolz, Thomas & Nataliya Levkovych. to appear. Function vs. form – On ways of telling repetition and reduplication apart. In Rita Finkbeiner & Ulrike Freywald (eds.), *Exact repetition in grammar and discourse*. Berlin/Boston: de Gruyter.

Stolz, Thomas, Cornelia Stroh & Aina Urdze. 2011. *Total reduplication. The areal linguistics of a potential universal*. Berlin: Akademie Verlag.

Wiese, Heike & Nilgin Tanis Polat. 2016. Pejoration in contact. *m*-reduplication and other examples from urban German. In Rita Finkbeiner, Jörg Meibauer & Heike Wiese (eds.), *Pejoration*, 243–267. Amsterdam: Benjamins.

Haritini Kallergi and Magdalene Konstantinidou

Reduplicative constructions involving distortion

An overview of Greek data with emphasis on the echo-word construction

Abstract: The paper gives an overview of Modern Greek reduplicative constructions characterized by non-identity between base and copy(ies), and in particular, by morphophonological variation in the copy(ies) (or distortion of the base) of various kinds. The cases examined include dialectal data, which sheds light into the formal features and semantic dimensions of distortive patterns, as well as their productivity and origin. Emphasis is placed on the echo-word construction (the *mühleme* or [m]-reduplication), as the Greek version of a widespread cross-linguistic phenomenon, as well as on a more local equivalent to this construction, the *X (ce) kse-X* superpattern. The discussion is led to observations about the commonalities between radically different patterns, under the rubric of "non-prototypical reduplication".

Keywords: Greek dialects, [m]-reduplication, echo-word construction, morphophonological distortion, pejorative

1 Introduction: prototypical vs. non-prototypical reduplication

According to the extensive study of reduplication by Stolz at al. (2011), reduplicative constructions resemble (more or less closely) a prototype, in the sense of a cognitively salient format of reduplication that is easier for the human mind to parse (Stolz et al. 2011: 45). This prototype or canonical form of reduplication (see Stolz, this volume) is defined by a number of principles (drawn from

Haritini Kallergi: Aristotle University of Thessaloniki, University Campus, Thessaloniki 54124, Greece. E-mail: haritinikallergi@gmail.com, harakallergi@yahoo.gr
Magdalene Konstantinidou: Academy of Athens (Research Centre for Modern Greek Dialects – I.L.N.E.), 22, Alexandrou Soutsou str., 10671 Athens, Greece.
E-mail: mkonstantinidou@academyofathens.gr

https://doi.org/10.1515/9783110599329-004

Mel'čuk 1996 and boils down to Total Reduplication of the *X X* type. More accurately, in its prototypical version, a reduplicative construction consists of two or more identical and adjacent syntactic words (Stolz et al. 2011: 34, 42), as in Modern Greek *komati komati* 'piece.ACC.SG piece.ACC.SG' "piece by piece".

But what falls outside this prototype? An answer may be "a vast number of different constructions where a word or part thereof reappears". In fact, the definition of non-prototypical reduplication depends on which aspect of prototypicality one is encountered with. One such aspect concerns wordhood: if the item that reappears does not fit our idea of a proper word, we are most probably faced with a case of non-prototypical reduplication, even though other formal/semiotic prerequisites may be met. For example, ideophones in Greek and other languages look like syntactic words and may appear as singletons (although more usually as reduplicative units of the *X X* type). However, they are less prototypical words, since they are combinations of a sound-symbolic/sound-imitative form with conventional meaning, rather than completely arbitrary form-meaning pairs with descriptive or grammatical meaning.[1]

In this paper, we are dealing with one large category of non-prototypical reduplication, namely reduplication involving formal distortion. That is, the second instance of the word that reduplicates (i.e., the copy or the *image*) is not identical to the reduplicand in that it appears as a variant thereof, carrying some different or extra phonological/morphological material.[2] The cases to be examined are the following: 1) reduplicative ideophones, in which, typically, the second part appears as a distorted copy of the first, as in *plits plats* 'water splashing all over'[3]; 2) fixed reduplicative expressions consisting of a possibly artificial word and a variant thereof. Typically, the variant carries a different

1 Actually, ideophones depict sensory aspects of events rather than describe aspects of events (see Dingemanse 2012). According to Dingemanse, this is what makes them "marked" cases of words, which for our purposes can be seen as synonymous to "non-prototypical".

2 Here, we take into consideration another important assumption about prototypical reduplication, i.e., that the copying occurs in a rightward direction (see Stolz et al. 2011).

3 The distortion here concerns typical ideophones that either come as doubles or singletons, such as *bam* 'sound of hitting or explosion', *plits* 'sound of liquid falling' and *taka taka* 'sound of repetitive hitting'. However, as we shall see, it is perhaps more correct to talk about reduplicative ideophone constructions with systematic vowel arrangement over the two parts of the dyad, i.e., either the first part has the form [consonant (cluster)] [a] [consonant (cluster)] and the second part has the form [consonant (cluster)] [u] [consonant (cluster)], i.e., CC[a]CC CC[u]CC, as in *frast frust* 'disturbing sound of hasty movement' or the construction has the form CC[i]CC CC[a]CC, as in *din dan* 'sound of bell ringing'. In other words, the case to be discussed here can be considered a type of *ablaut reduplication* in Modern Greek (i.e., similar to a type of reduplication exemplified by *riff-raff* and *chit-chat* in English, see e.g., Minkova 2002).

first consonant (cluster), as in *tsatra patra* 'in an unorganized and sloppy manner', and 3) productive constructions involving reduplication with some kind of distortion. Emphasis will be placed on one of them, the echo-word construction or *mühleme*, whereby, typically, the second instance of a word carries a [m] initial consonant, as in *ozes mozes* 'nail.enamel.PL ECHO' "nail enamels and the like" (see, e.g., Konstantinidou 2004). The rest of cases that fall into the third category do not involve distortion via a phoneme, but via the addition of a fixed morpheme. Almost all of these morphemes have the form of existing prefixes in Modern Greek: *kse-* (roughly 'un-' or de-'), *para-* 'para-', *andi-* 'anti-' and the marginal (in fact, slang), *skat-* 'shit-'.

In all of the above cases, various conditions for prototypicality à la Stolz et al. (2011) are not met. First, the distortion immediately cancels the critical criterion for Total Reduplication, i.e., exactness in copying. Also, in many of these constructions the two instances of reduplication are interrupted by *ce* 'and', so that they are not immediately adjacent. Most importantly, either one or both parts of the reduplicative dyad are non-words.[4] This makes them particularly interesting as a category, because the "non-wordhood" of the copy, as well as the morphophonological distortion involved, seems to relate to particular semantic and pragmatic effects across these practically different construction types. More concretely, there are interesting functional parallels between patterns of distortion. Also, there are probably iconic connections between formal distortion and semantic/pragmatic effects (see, e.g., Konstantinidou 2004, Stolz 2008, Kallergi 2013). Thus, in parallel to discussions of the functional core of prototypical, total reduplication, it would be interesting to explore whether there are any functional characteristics of non-prototypical reduplication, at least of the type discussed here.

The discussion is structured as follows: The three types of non-prototypical reduplication mentioned above are examined one by one in Section 2. From the third type, echo-word constructions and *X kse-X* constructions are discussed in separate subsections (2.3.1 and 2.3.2 respectively). The rest of constructions are

---

4 We use this term to refer to three cases: i) ideophones, in the sense of them being distinct from proper words, as explained above; ii) artificial words, i.e., word-forms that do not exist in the language with a specified meaning. It might be the case that they are semantically opaque words, but extensive etymological study is needed to prove this; and iii) word-forms that do not carry a meaning of their own in the lexicon and are not met outside the reduplicative constructions in question. Since these word-forms are often the result of the distortion by the addition (or replacement of segments by) existent prefixal forms, they may sometimes be identical to existent words. However, this effect is accidental.

referred to without much detail in Section 2.3.3, since they seem to be relatively minor with respect to frequency in Modern Greek.

Note that the Modern Greek data used in this paper include dialectal speech. In fact, there are three types of data that come from the following sources: i) data on Standard Modern Greek (SMG onwards) that come from grammars and dictionaries of the language as well as from personal records of the authors: these include data from everyday conversations in which we were present, Google hits on specific examples, and records of informal language in theatre and the media; ii) dialectal data, which comes from fieldwork by Magdalene Konstantinidou, particularly interviews with elderly people from Bithynia (Prusa/Bursa) and other parts of Asia Minor (see Konstantinidou 2004, 2005), and iii) some data (dialectal or not), which come from resources of the Academy of Athens, particularly ILNE (the abbreviation standing for the *Historical Dictionary of Modern Greek, both of the common spoken language and the dialects*, vol. 1–6) and the *Archive of the Research Centre for Modern Greek Dialects – ILNE* (Archive ILNE onwards).[5]

2 An overview of reduplicative constructions involving distortion

An important distinction to be emphasized at this point is that Modern Greek (MG) has both fixed and productive reduplicative constructions with variation at its disposal. The type discussed in 2.2 certainly resembles non-productive reduplication types in English and other languages (particularly rhyming reduplication, such as *teenie-weenie*, and ablaut reduplication, such as *chit-chat*). However, reduplicative ideophones involving distortion (or ablaut processes, for the most part) cannot be included as a category of fixed expressions. Speakers are highly creative with ideophones, and although they have a more or less specific inventory of ideophones at their disposal, they may present considerable variation as to which ones they use most frequently, as well as to which ones they would use on a particular occasion (viz., which exact form they prefer, choosing from a great range of possibilities). Individual variation is particularly evident in ideophones with variation/distortion.[6] For this reason, the first type

5 These data are mainly the result of fieldwork taken place during the last as well as the current century by researchers of the Academy of Athens and other scholars in different dialectal areas of Modern Greek (see Konstantinidou & Tzamali 2012 for details).

6 To put it differently, this kind of ideophones seems to present "individual productivity", rather than "societal productivity" (Bauer 1988: 65).

to be examined (Section 2.1) will be considered a semi-productive category of reduplication with variation/distortion.

2.1 Reduplicative ideophones involving variation/distortion

Ideophones, as viewed here, are a type of sound-symbolic units, i.e., units that directly connect sound with meaning, particularly bodily or mental states or even states of affairs (see Hinton et al. 1994). Ideophones may be used to form words (e.g. MG *tsirtsir-izo* 'sound of sizzling/frying/dripping-1SG.PR' "to sizzle") or may be used as such to fulfill syntactic roles in sentences (mostly having adverbial or nominal use, as in *to treno ekane tsaf-tsuf* 'the train made tsaf-tsuf' "the train went tsaf-tsuf", see Joseph & Philippaki-Warburton 1987).

The ideophones mentioned in Table 1 below may also be said to form two categories: i) sound-imitative ideophones, such as *plits-plats* 'sound of water spilt all over', *krits-krats* 'crispy sound of something being broken', *tsaf-tsuf* 'sound of train engine' etc. and ii) ideophones representing a state of affairs or a state of mind (or even, a stance, Joseph & Philippaki-Warburton 1987), such as *tsuku-tsuku* 'slowly', *tak-tuk* 'fast, hastily', etc.

Table 1: Parallels between reduplicative ideophones involving identical constituents and reduplicative ideophones involving morphophonological variation.

X X ideophones		**X Y ideophones**	
bam bam	shooting	*bam bum*	noise of hitting, fuss
bum bum	hitting	*baka buka*	intense hitting/shooting
		daba duba	noise made of drums
plits plits	dripping	*plits plats*	spilling water
		plats(a) pluts(a)	spilling water all over
tik tik	ticking	*tik tak*	heart/clock beating
tak tak	repetitive sound	*tak(a) tuk(a)*	quickly, hastily
guxu guxu	coughing	*gaxa guxa*	nasty cough
drin drin	phone ringing	*dranga drunga*	disturbing noise
krits krits	repetitive crispy sound	*krits krats*	intense chewing/crushing
krats krats	crushing		
tsuku tsuku	moving slowly (and with difficulty)	*tsak(a) tsuk(a)*	fast, hastily (with ease)
		mats muts	(ludic) kiss
		frast frust	hastily, disturbing sound
din din	(small) bell ringing	*din dan*	church bell ringing
		din don	house bell ringing

Note, also, that the interpretations given here are not at all exhaustive of the dynamics of the ideophones in question, but they will do for the purpose of showing a kind of equivalence across ideophones involving total reduplication and those involving variation.

A few notes on phonology are in order. It seems that there is a systematic vowel alternation across the ideophones of the second column, in that (i) most of them have the structure C[a]C C[u]C (where C may be a consonant or consonant cluster), and (ii) the rest are formed along the C[i]C C[a]C (or, less often, C[i]C C[o]C) template, (very similarly to the English ablaut reduplication, e.g. *flim flam*, *hippety hoppety*). Also, concerning the second category only, an additional final vowel [a] may appear, but its presence seems to be optional (e.g., *plats(a) pluts(a)*), except a few cases (e.g., *baka buka*, *daba duba*).

Concerning the connection between these phonological features with interpretation in ideophones with variation/distortion, one might observe that the examples belonging to category (i) above express greater intensity than those of category (ii). That is, the interpretation of *plats(a) pluts(a)* 'spilling water all over' seems to definitely involve the feature "all over", in relation to its peer with the [i] > [a] structure (*plits plats*). Also, the cases with the [a]>[u] structure perhaps sound more informal (which may be viewed as a concomitant of the fact that they express greater intensity). The effect of higher informality and intensity seems to be reinforced by the presence of the final [a]. It is also likely that this final vowel adds to a sense of playfulness/humorousness with respect to the ideophone that lacks it (here, *plats pluts*).

Concerning the meaning differences across the two columns, a general remark may be that the emotions are expressed more by the ideophones of the second column than the ones of the first column (if at all). The latter tend to describe states of affairs, whereas the former tend to bear implications about these states of affairs (see, e.g., the "unmatched" cases across the table that appear only with a variated structure, or compare, e.g., *guxu guxu* 'cough' and *gaxa guxa* 'nasty cough'). Specifically, ideophones without variation usually express (repetitive) sounds (e.g., *drin drin* 'phone ringing'), whereas ideophones involving distortion tend to express the emotions evoked by the repetition or the state of affairs itself (e.g., *dranga drunga* 'annoying repetitive sound' (as, e.g., of a badly played guitar)). In other words, the variation or distortion seems to add an expressive component on the semantics of a reduplicative ideophone.

Regarding semantics, the overall meanings of the ideophones involving distortion seem to evolve around notions such as intensity, razzmatazz, lack of order and lack of control. In particular, greater intensity is translated to louder sound (e.g., *dranga drunga*) or higher speed (e.g., *tak(a) tuk(a)*) or something relating to bigger size (compare *din din* and *din dan/don* and their implications on the size of bell). Razzmatazz is expressed nicely in *bam bum*, *baka buka* ('intense sound of hitting/shooting, fuss') and *daba duba* ('noise made of drums'). Also, a combination of intensity and lack of order/control may be said to be reflected in those cases that mean "quickly, hastily", such as *tsak(a) tsuk(a)*, *tak(a) tuk(a)*, and *frast frust*. Finally, in *platsa plutsa* 'water spilled all over', we see a typical example of lack of control, in the sense of continuity or organization: spilling can only be discontinuous and uncontrolled, whereas dripping (*plits plits*) involves natural continuity. In general, repetition (as expressed by prototypical reduplicative ideophones in the first column) involves continuity in time and space, which seems to be disrupted in the case of non-prototypical reduplicative ideophones.

With respect to pragmatics, it is evident from the table that X Y ideophones generally have negative implications and/or express negative emotions. The discontinuity that comes as a direct effect of distortion or variation seems to evoke feelings of annoyance: intensity is rather negatively appreciated in this case, whereas lack of order, control and organization is generally deemed unfavorable. However, as ideophones in general, X Y ideophones retain their character as elements of child language or vivid narrative speech. As such, the negativity implied by the semantics of these ideophones is shaded by pragmatic implications of playfulness, intimacy and humor. These features are particularly evident in cases such as *mats muts* '(intense) kiss' and *platsa plutsa* 'water spilling/spilled all over'. Note again that these examples relate to highly informal speech situations. In these contexts, it is easy to move from playfulness and humor to irony and contempt; this is especially evident in instances involving the final [a], such as *dranga drunga* and *daba duba*, which can hardly appear in non-intimate speech situations.

In sum, the semi-productive mechanism of using reduplication with phonological variation/distortion yields a category (or categories) of ideophones, which are generally more expressive than their peers involving prototypical (total) reduplication. Their semantics is oriented towards greater intensity and discontinuity (whether this means lack of order, control or organization). These effects seem to iconically connect the discontinuity in reduplication (effected by the variation) with discontinuity in meaning. Also, X Y ideophones express additional informality, intimacy and playfulness or vividness with respect to X X ideophones. In many of these cases, however, the semantics of X Y ideophones may have negative implications, such as annoyance and contempt.

2.2 Fixed reduplicative expressions involving consonantal variation/distortion

In 2.1, we examined reduplicative ideophones characterized by phonological variation with respect to their constituent vowel(s). This variation may be viewed as distortion if we consider ideophones with identical constituents (i.e., of the X X pattern), that are based on the same consonants/consonant clusters with the X Y ones. In this section, we look at completely unproductive (frozen) expressions made up of two non-words (see fn. 4), whereby the variation concerns the first consonant or consonant cluster of the second constituent.

Table 2: Fixed expressions involving reduplication with consonantal variation.

EXPRESSION	MEANING
1. *ares mares (kukunares)*	incomprehensible speech, nonsense
2. *gir mir* (Northern Greece)[7]/*gix mix*	mubbling, nagging, complaint
3. *suksu muksu*	unclear/hidden talk, gossip, conspiracy, hanky-panky
4. *saʎa baʎa* (Mytilene)	nonsense, something unorganized, a mix of non-matching things
5. *saxla maxla* (Epirus, Thrace)	see number 4
6. *artzi burtzi*	nonsense, something completely lacking (logical) order
7. *tsatra patra*	in an unorganized, sloppy manner
8. *firðin miɣðin*	all mixed up, in complete disorder, topsy-turvy
9. *alaθra malaθra* (Pontus)	see number 8
10. *dari-mari* (Euboia)	see number 8
11. *dzigri-migri* (Aegean, Santorini)	see number 8
12. *dzidzi(li) midzi(li)*	host of different small things
13. *dzadzala madzala*	group of undefined, unimportant things
14. *i sara ce i mara*	(group of) people of all kinds [+negative]
15. *to suri ce to muri* (Euboia)	see number 14
16. *arades marades (xarxulivades)* (Aegean, Lesbos)	nonsense

As is evident from the examples in Table 2, the consonant in question is very often [m]. Alternatively, [p], [b] and [g] appear as the onset consonants of the

7 Cf. Albanian *gerr-merr* 'nagging' (Stolz 2008: 118, Map 1).

second constituent, which reveals a principle of preference of either a bilabial or plosive consonant. Other consonants (mostly alveolar, such as [r], [l], [n], [s]) may appear occasionally and without systematicity. For example, along with SMG *ares mares* 'nonsense', the dialectal instances *ares p̲ares* (Pontus), *ares b̲ares* (Euboea), *ares n̲ares* (Propontis) are also attested (Konstantinidou 2004: 351). Other MG dialectal instances (taken from the Archive ILNE and Konstantinidou 2005: 267) include *atala batala* 'nonsense' (Epirus), *surdu burdu* 'higgledy-piggledy' (Aegean, Karpathos), *ada rada* 'topsy-turvy' (Northern Greece, Thessaloniki), *sima lima* 'hotchpotch, farrago' (Central Greece, Thessaly), *arakin sarakin*, *puria suria* (incomprehensible words from an incantation, Southern Greece, Peloponnese). Cf. also Medieval Greek *satala patala* 'incoherent, unintelligible/incomprehensible words, nonsense' (13th cent., manuscript 16th cent., see Kriaras 1968, s.v. σάταλα). For similar observations in Georgian, see Kikvidze (2016).

It is important to note that for some of the items in Table 2 there are assumptions on etymology and origin in dictionaries.[8] Other items are only dialectal (see Archive ILNE, Politis 1900: 415). Also, the list is not exhaustive, but only indicative of a large number of like expressions. Some more attestations reveal a wide distribution across Greece (or formerly Greek-speaking regions): *surðu murðu* 'in complete disorder, topsy-turvy' (Northwestern Greece: Epirus, Ionian: Cefalonia, Lefkada, Southern Greece: Peloponnese) (Archive ILNE, s.v. *σούρδου-μούρδου*), *dzigri migri* 'all mixed up, topsy-turvy' (Aegean: Santorini) (Archive ILNE, s.v. *τζίγκρι-μίγκρι*), *tsaɣði maɣði* 'all mixed up, in complete disorder, topsy-turvy' (Pontus) (Papadopoulos 1958, s.v. *τσάγδη*), *tsicir micir* 'knickknacks, bric-a-brac' (Northwestern Greece: Epirus) (Archive ILNE, s.v. *τσίκιρ-μίκιρ*),

Such expressions are found in dialectal monumental speech, i.e., riddles, nursery rhymes, counting-out games and tongue twisters. The frequency of such expressions in these types of text implies that the process seems to have been productive at earlier stages of the language. In these genres, a word and a

---

8 These are examples 6 and 8 in Table 2. For instance, both constituents of the expression *firðin miɣðin* are assumed to originate in earlier stages of Greek (Ancient Greek adverbs *φύρδην* and *μίγδην*). If this is correct, then this expression should not belong to the type of reduplication under discussion, since it comprises a combination of two distinct words. However, there are attestations of the expression in the form *firðin mir̲ðin* (see, e.g., www.farassiotis.gr/kappadokia/history/agiosarsenios.htm), in which case the second constituent is assumed to be a reduplication of the first according to the X [m]X pattern (for more see Konstantinidou 2004 and the *Pandora* journal dated 1850 – *Πανδώρα* 1(14): 537). Similarly, *artzi-burtzi* is reported in many SMG dictionaries to originate in a single word of Armenian origin. Thus, despite its frequency or popularity, it is marginally a case of interest here.

distorted copy thereof may also appear in (syntactic) distance between each other and/or in various combinations between each other, as, e.g., in (1):

(1) Tongue twister, Argos, Peloponnese [*Laographia* 2, 1910: 203]
O dzindziras, o mindziras, o dzindzimindzixondziras...
the cricket [old form] the ECHO the cricket-ECHO

Thus, the frozen expressions in Table 2 seem to be the lexicalized results of a broader mechanism of using a word and its variation for several semantic and pragmatic purposes.

Regarding the semantics of Table 2, one may observe three broad – yet interrelated – categories of meaning: (1) 'incomprehensible speech/nonsense', (2) 'something lacking order/organization', and (3) 'group of undefined (and negatively marked) things'.

With respect to the first category (expressions 1–6), one may identify the meaning "nonsense" on the one hand, and the element "incomprehensible" or "hidden" on the other. Especially the latter semantic element seems to relate to the most probable origin of such expressions, viz., magic and cryptic language (see 2.3.1.6 below). More specifically, expressions of this kind abide in children's verbal games, whereby they often have a "cryptic, puzzling function" (Konstantinidou 2004: 350). For instance, in the following examples the artificial words (the second being a copy that starts with [m]) have no meaning and typically signify the hidden word/object of the puzzle:

(2) Chili/Sile, Bithynia [Hatzitaki-Kapsomenou 2001: 330]
ðos me to sidri su, to midri su/ na sidri-so,
give me the sindri yours the ECHO yours/ to sindriso-1SG
na midri-so...
to ECHO-1SG

This confusion-provoking function of the pattern is especially found in MG dialectal riddles. In all of the following dialectal instances, the reduplicative expression has no meaning but simply replaces (and in this way camouflages) the words that represent the solution of the riddle: *segles megles, serðe merðe, sidri midri* (Asia Minor: Chili Bithynia), *sirði mirði, sidi midi, tsala mala* (Northeastern Greece: Thrace), *tsindzili mindzili* (Northern Greece: Serres, Northwestern Greece: Kozani), *tsendzilo mendzilo* (Northwestern Greece: Epirus), *sfigina migina, tsidzilo midzilo* (Central Greece: Thessaly: Karditsa), *to tsindzelo to mindzelo* (Central Greece: Evritania: Granitsa), *sede mede* (Southern Greece: Peloponnese, Aegean: Crete), *tsadzala madzala, tsidzina midzina, tsindzi mindzi* (Central Greece: Euboea), *sendena mendena* (Eastern Aegean: Chios), *sarðes marðes* (Eastern Aegean: Samos), *indina mindina* (Aegean: Skiathos), *sendzi mendzi* (Aegean: Skyros), *sayðe mayðe* (Aegean: Cyclades: Naxos), *segli megli* (Aegean: Cyclades:Tinos), *serðe merðe* (Southeastern Aegean: Dodecanese: Syme) (Hatzitaki-Kapsomenou 2001, Archive ILNE).

Regarding the second category (7–11), lack of order and organization comes in similar terms to the meaning of several X Y ideophones discussed in the previous section: the negative emotions implied may combine with shades of humor and playfulness. Playfulness is again reminiscent of the function that the pattern has in children's counting games, where funny-sounding (and semantically empty) numerals are used:

(3) *Ai-vai/come stai/die-mie/companie/sami-rako/tici-tako/puf*
[Ithaca, Archive ILNE, s.v. *ντίε μίε*]

(4) *Trigi, pegi, ligir, tutli, mutli...* [Cappadocia, Anastasiadis 1976: 5]

Other relevant examples are *sendeli mendeli* (Pontus), *aretsi maretsi* (Asia Minor: Bithynia), *karela marela* (Southern Greece: Arcadia Peloponnese), *suiđu muiđu* (Northern Greece: Halastra Thessaloniki), *surðu murðu* (Aegean: Chios) (Konstantinidou 2004: 350, Konstantinidou 2005: 264, Anastasiadis 1976: 5, Archive ILNE). Here, (repetitive and rhyming) sound seems to play an important role. However, it is also possible that the use of the pattern in counting also relates to the concept of grouping, which brings us to the third category.

Expressions 12–15 on the table carry the meaning "group of undefined/unspecified things" plus a negative connotation. As we shall see in 2.3.1 below, the juxtaposition of an actual word to its symbolic copy yields the meaning "X and the like" (=group of similar things) in many languages. Here, the juxtaposition of an artificial word to its (bad, i.e., distorted) copy has strong implications on the quality and importance of this group of things; they are small, unimportant or useless and negatively inhomogeneous.[9]

In sum, a common feature between these meaning categories is the element of lack of order or lack of homogeneity, which seems to boil down to the element of discontinuity, as was observed for X Y ideophones in 2.1. When the reduplicative expression does not carry meanings of this kind (i.e., 'nonsense', 'topsy-turvy', 'inhomogeneous group of unspecified things'), and, in fact, when it does not carry meaning at all, reduplicative distortion seems to serve the func-

9 Note that the word *dzadzala* in expression 13 in Table 2 has medieval origin and seems to already mean "small, unimportant things", according to the dictionary of Kriaras 1968 (see the word *ἄταλα/ἄντζαλα atala/andzala* there).

tions of concealment and funniness/playfulness, particularly observed in children's verbal games (esp. in dialectal speech). The latter functions seem to connect to the possible origin of distortive reduplicative patterns in (verbal) magic (see 2.3.1.5.2 below).

Similar meanings arise from rhyming reduplication in English and other languages (including German, see, e.g., *Heckmeck* 'nonsense, rubbish', *Kuddelmuddel* 'mess-up' etc., Müller 2004: 52 n. 164, 327). For instance, the English frozen expressions *hocus-pocus* 'nonsense or sham', *topsy-turvy* 'in complete disorder' and *razzle-dazzle* 'state of confusion or hilarity' encompass most of the meaning possibilities of the mechanism across Greek and English. So far, that is, different patterns of distortion (phonological, morphological and variations thereof) seem to result in very similar semantic and pragmatic effects. In addition, similar formal patterns of non-prototypical reduplication seem to share similar meanings across languages. The only pattern that does not seem to abide in English (but rather tends to appear in Eastern languages) is echo-reduplication with the substitution of the onset consonant (cluster) of the reduplicated word with [m] (as in most expressions in Table 2). In its productive version, this mechanism is closely examined in 2.3.1 below.

2.3 Productive reduplication with fixed segments in Modern Greek

As mentioned in the introduction, productive reduplication that involves variation/distortion in Greek comes in two large categories: the first, known as *echo-word construction* in the international literature, involves the addition of or substitution by a phoneme (typically, [m]) (Section 2.3.1); the second involves the addition of or substitution by a fixed morpheme, which almost always has the form of a productive prefix in MG (Section 2.3.2).

2.3.1 The echo-word construction (X [m]X or *mühleme*) in Modern Greek

In this subsection, we discuss another type of "total reduplication-cum-variation" (TRCV) in Modern Greek, namely the so-called echo-word formation (Stolz 2008) or *distortive reduplication with [m]* (Konstantinidou 2004, 2005), known also as *mühleme* (in Stolz 2008: 116, taken from Johanson 2002: 81). The pattern has primarily been described in dialectal Modern Greek (in a dialect of Bithynia, Asia Minor) in Konstantinidou (2004) and (2005). Brief discussion of the phenomenon, mainly within the framework of general grammatical/syntax descriptions of MG local varieties (dialects of Pontus and Cappadocia) or etymological studies on fixed reduplicative patterns or certain unusual numerals containing *mühleme*, is to be found in Valavanis (1892), Psaltes (1915: 109), Dawkins (1916: 117–119), Anastasiadis (1976: 289), Athanasiadis (1977: 32). Some semantic-pragmatic parameters of the pattern are also discussed in Kallergi (2013). With regard to SMG there is so far little data available, mainly derived from daily conversations, TV/theatre speech, or media language (see Thomopoulos 1986: 735, Konstantinidou 2004: 352 and 2005: 267, Sarantakos 2013). There are also two relevant lexicon entries (s.v. μ- [m-]) in some dictionaries for SMG, i.e., DSMG 1998 and UMGD 2014. A specific study on SMG *mühleme*-pattern is nevertheless a desideratum.

2.3.1.1 Phonology of Modern Greek *mühleme*

The Modern Greek *mühleme* displays the following phonological features: The bilabial nasal [m] replaces a consonant or consonant cluster at the onset of the first syllable of the reduplicand. For instance:

(5) dialectal, Crete [Konstantinidou 2004: 349]
strigla *migla*
bitch ECHO

If the word-initial syllable is empty, then [m] is added. For example:

(6) dialectal, Northern Greece [H. Kallergi, personal archive]
ozes *mozes*
nail.enamels ECHO

When the word-initial syllable and the word-second syllable are identical, [m] replaces the consonant or consonant cluster at the onset both of the first and of the second syllable of the reduplicand. For example:

(7) dialectal, Bithynia [Konstantinidou 2004: 348]
bebeko *memeko*
bebeco ECHO

If the base word begins with [m], *mühleme*-constructions are avoided or other types of TRCV are selected, e.g., the base word is reduplicated with vowel alteration, of the type seen in reduplicative ideophones involving variation/distortion (see 2.1. above). For example:

(8) SMG [Babiniotis 1998, s.v. *μα* 1, UDMG, s.v. *μα* 1]
ma *(ce)* *mu*[10]
but (and) ECHO
'objections, waverings, pretexts [+informal, +negative]'

A possible dialectal instance of this kind is suggested in Athanasiadis (1977) in the following example, where the base word *Martis* 'March' is reduplicated with vowel alteration [a] > [u].[11]

(9) Pontic dialect [Athanasiadis 1977: 32]
Martis *Murtis*
March ECHO
'evil March'/'March Schmarch'

Another way of treating cases with [m]-initial form within the framework of the *mühleme*-pattern is suggested in (10)–(11), where the echo-form is followed for extra emphasis by a second echo-form, which is a copy of the first echo-form starting with the dental (or alveolar) fricative [s]. It is interesting that the same pattern is already to be found in a vernacular Medieval Greek text (14th–15th cent.) (see example 12, cf. also Latin case in 52).

(10) SMG [Konstantinidou 2004: 352]
Citakse! *ðen* *eçi* *IKA,* *MIKA,* *SIKA!*
look.IMP not have.3SG.PR IKA ECHO ECHO
'Look! There's no IKA[12] and such things!'

(11) SMG[13]
i *Norvijia* *ðen* *eçi* *IKA,* *MIKA,* *SIKA,*
the Norway not have.3SG.PRS IKA, ECHO ECHO

10 Cf. synonym *ma (ce) kse-ma* in 2.3.2.

11 Note that *Murtis* is exclusively attested in a pontic proverb about March (*Martis Murtis, ajelastos, kseroxalxanizmenos* 'March Schmarch, sullen, giggling disgustingly', also in the variation *Marti m', Murti m', ajelaste ce kseroxalxaniste* 'My March, my Schmarch, you sullen and being giggle disgustingly', Archive ILNE). Papadopoulos (1958: s.v. *μούρτης*) connects *murtis (μούρτης)* with *murða (μούρδα)* 'dregs, sediment' (< Ancient Greek *ἀμόλγη*), and considers it as an adjective, meaning 'turbid, fig, cloudy'. Nevertheless, the absence of any other evidence of *murtis* beyond the mentioned proverb, in combination with the data of the above-mentioned variation which indicates that *Murtis* shares the same word class with *Martis* (i.e. that of a proper name), suggests that *Martis Murtis* should be seen as a case of the [m]-reduplicative pattern.

12 An acronym of a Greek insurance fund.

13 At www.patriotaki.net/μετανάστευση-στην-ευρώπη-96/.../index27.html (checked 14/09/2012). The pattern is based on the acronyms of two insurance funds (IKA and TEVE).

TEVE,	*MEVE,*	*SEVE*	*ala*	*mono*	*ena*	*tamio!*
TEVE	ECHO	ECHO	but	only	one	insurance.fund

'In Norway there is no IKA and such things, no TEVE and such things, but only one insurance fund!'

(12) Medieval Greek
[Eideneier, *Spanos* 1977 B 49–50, MS -*ντσ*-, Kriaras, *Epitome* s.v. *άταλα* 2]

(μάζωξε)	*ἄντζαλα*	*μάντζαλα*	*σάντζαλα*
collect.IMP	non.word.N.PL	ECHO	ECHO

'(collect) andzala mandzala sandzala' (i.e., 'unimportant or inexistent things', Kriaras, *Epitome* s.v. *άταλα* 2)

In some cases, such as those exemplified in (13)–(14), the m-reduplication is accompanied by vowel alternation as well, e.g., [i] > [a] or [a] > [u], and this again shows similar systematicity with reduplicative ideophones involving variation/distortion (2.1 of this study).[14] Whether this double phonological variation has a reinforcing effect on the meaning of the reduplicative pattern cannot be answered with certainty.

(13) dialectal, Bithynia [Konstantinidou 2004: 347]

tsifuti	*mafuti*
stingy.MASC	ECHO

'stingy, schmingy'

(14) Pontic dialect [Athanasiadis 1977: 32]

axara	*muxara* (besides *axara maxara*)
gracelessly	ECHO

'in great difficulty'

Note also that Valavanis (1892: 60) refers to a pattern of echo-word formation with [p] as initial segment of the reduplicand in the MG dialect of Pontus (Asia Minor) along the *mühleme*-pattern, yet he does not provide any example or other evidence thereof, in order for one to decide whether this TRCV case forms a productive pattern in the Pontic dialect. Similar patterns can be identified in Turkish (Müller 2004: 54–55) as well as Armenian and Hungarian (Stolz 2008). In yet other languages, variation/distortion with the use of consonants other than [m] (e.g., [g]) and vowel alternation seems to be the norm (see, e.g., Keane 2005).

14 Some other phonological alternations that may be attested on the copy word are of merely dialectal nature, e.g. the alternation [f] > [v] in the Cypriot pattern *stafiliti maviliti* vs. *stafiliti mafiliti* (both *mühleme*-constructions are encountered in charms against *stafilitis* 'uvulitis, a disease of the throat', Cypriotic Greek, Archive ILNE, s.v. *μαβιλίτης*, Passalis (2012: 12).

2.3.1.2 Morphology of Modern Greek *mühleme*

The m-initial echo-word formation pattern is productive in MG mostly with nouns (both common and proper nouns) and adjectives (examples 15–17). It rarely appears with pronouns, numerals, verbs, adverbs or interjections, (examples 18–22).

(15) noun: *strigla migla* 'bitch ECHO'
[dialectal, Konstantinidou 2004: 349]

(16) proper noun: *θanasis Manasis* 'Thanasis ECHO'
[dialectal, Konstantinidou 2004: 347]

(17) adjective: *anatomika manatomika* 'anatomical.PL ECHO'
[dialectal, Konstantinidou 2004: 348]

(18) pronoun: *kanis manis* 'none ECHO'
[dialectal, in a document dated 1815, Loukopoulos 1938: 447]

(19) numeral: *saranda ce maranda* 'forty and ECHO'
[dialectal, Konstantinidou 2004: 346]

(20) verb: *griɲazes miɲazes* 'nag.2SG.IPFV.PST ECHO'
[dialectal, Kallergi 2013: 319]

(21) adverb: *ekso mekso* 'out/outside ECHO' [SMG, DSMG, s.v. *μ*-]

(22) interjection: *puf muf* 'alas ECHO'
[dialectal, Byzantios' *Babylonia* B 41, Evangelatos 2002]

A few lexicalized cases of m-reduplicative patterns may be traced in SMG (Konstantinidou 2004: 351), e.g., (23)–(24). In (24) the initial consonant of the base word is substituted by [r] instead of [m].

(23) adverb *ksana* 'again' → *ksana-mana* again ECHO

(24) verb *kano* 'do' → *kano-rano* do ECHO

The example under (8) above may be also considered as a lexicalized case of a *mühleme*-construction in SMG, i.e., conjunction *ma* 'but' → *ma (ce) mu* 'but (and) ECHO' "objections, waverings, pretexts [+informal, +negative]" (e.g., *ðen eçi ma ce mu* 'objections are not allowed!' – see example 68).

Interestingly, in dialectal MG there are cases like (25), in which the frozen expressions involving distortion with m-initial show productivity, by forming derivatives either from the phrase as a whole or from one of the two constituents of it, that is, they behave as ordinary autonomous words, (e.g., 25–26):

(25) Northern Greece: Thessaloniki, Halastra, Northwestern Greece: Epirus, Konitsa [Archive ILNE, s.v. *σούιδου, σούϊδου-μούιδος*]

súiðu *múiðu*
[non-word] ECHO
'nonsense words representing numerals in children's games or tongue-twisters'

→ *(ise) súiðu-múið-os*
be.2SG non.word-ECHO-MASC
'(You are) very same, exactly the same, selfsame, counterpart of someone'

(26) Northwestern Greece: Epirus
[Archive ILNE, s.v. *μπάνταλα, μπανταλός, μπανταλομάρα*]

adala	*badala* →	*badal-os*→	*badal-o-mara*
[non-word] ECHO		ECHO-MASC	ECHO-/o/-N.FEM
'nonsense'		'fool, stupid'	'stupidity'

The *mühleme*-pattern often combines with plural, i.e., both base word (X) and reduplicand ([m]X) are used in plural form. This is predominantly the case when the base of the construction is a proper name, (as, e.g., in 27), and occasionally when it is a common noun without plural form (Konstantinidou 2004: 346). By this means the basic meaning of the pattern is reinforced. Note that plural alone can be used to express negativity in MG, especially when it is not predicted (e.g., with proper names) or is truth-conditionally false.

(27) dialectal, Bithynia [Konstantinidou 2004: 346]

emis	*eprepe*	*na*	*jinume*	*estiatores!*	*Vef-es,*
we	should.PST	PTCL	become.3PL	restaurant.owners	Vefa-PL
Mefes	*ðe θa*	*pçanane*		*tipote!*	
ECHO	not FUT	catch.3SG.IPFV.PST		nothing	

'We should have become restaurant owners! Vefa and such chefs would not count at all!'

2.3.1.3 Syntax of Modern Greek *mühleme*

The base word (X) and copy ([m]X) are sometimes connected conjunctively or disjunctively with *ce* 'and', or with the *ce X ce Y* ('both X and Y'), *ute X ute Y* ('neither X nor Y'), *ite X ite Y* ('either X or Y') constructions (examples 28–31), but not when serving a concessive function (Kallergi 2013: 322).

(28) SMG [DSMG, s.v. *μ*-]

Exo		*vareθi*	*na*	*vlepo*		
have.1SG.AUX		bore.INF	PTCL	see.1SG		
kaθe	*toso*	*ta*	*soja*	*ce*	*ta*	*moja*
every	this.much	the	kindred.ACC.PL	and	the	ECHO

'I am very bored facing every now and again the kindred and the schmindred'

(29) SMG [TV comedy series *ðio kseni*, MEGA Channel, 1997–1999]

Na	*pane*	*sto*	*ðjavolo*	*ce*	*i*	*Cislofsci-ðes*
PTCL	go.3PL	to.the	devil.ACC	und	the	Kieslowski-PL
ce	*i*	*Mislofsci-ðes*	*ce...*	*ce*	*ola...*	*ola*
and	the	ECHO-PL	and	and	everything	everything

'To hell both Kieslowski and Schmieslowski and... and... everything!'

(30) dialectal, Bithynia [Konstantinidou 2004: 346]

(ðen	*irθe)*	*ute*	*i*	*Litsa*	*ute*	*i*	*Mitsa*
(not	come.3SG.PST)	neither	the	Litsa	nor	the	ECHO
ute	*kanis*						
nor	anybody						

'Nobody came, neither Litsa (nor Schmitsa) or anybody else'

(31) SMG [Alpha radio-journal, 28/06/2016]

ðen	*bori*	*o*	*opçosðipote*	*ite*	*lejete*	*Sulz*
not	can.3SG	the	anybody	whether	be.named.3SG	Schulz
ite	*lejete*	*Mulz*	*na*	*lei*	*tetça*	*pramata*
or	be.named.3SG	ECHO	PTCL	say.3SG	such	thing.PL

'No one should say such things, whether he is named Schulz or Schmulz'

It is worth noting that occasionally the copy component of a *mühleme*-construction is not immediately adjacent to the original one, but tends to behave autonomously as a syntactic word (always of the same part of speech with the base), (see example 31). The same scope extension from word to clause is also attested in the following example (32), which makes use of an expression that has been mentioned above (under 24) as a rather fixed/lexicalized case (with substitution by [r] rather than [m]):

(32) SMG

na	*mi*	*se*	*ɲazi*	*ti*	*kan-o*
PTCL	not	you.ACC	bother.3SG.PRS	what	do-1SG.PRS
ce	*ti*	*ran-o!*			
and	what	ECHO-1SG.PRS			

'It's none of your business what the hell I do!'

An issue for further study would be to determine all the possible syntactic roles/functions of the *mühleme*-constructions as part of a sentence.

2.3.1.4 Distribution and frequency of Modern Greek *mühleme*

Regarding SMG, the *mühleme* seems to represent a productive but not frequent reduplicative pattern. See, e.g., *ekso mekso* based on *ekso* 'outside', *soja moja* based on *soja* 'relatives' (DSMG, s.v. *μ*-), *portes mortes* door.PL ECHO, *solines molines* tubing.PL ECHO (Thomopoulos 1986: 735), *IKA MIKA (SIKA)* based on the acronym for Social Insurance Institution *IKA* (UMGD, s.v. *μ*-), *Tsamberlen Mamberlen* based on *Chamberlain* (Greek movie: A. Sakellarios, 1960, *Makrikostei ce Kondojorjiðes*), *Juncer Muncer* based on surname *Juncker* (*The National Herald GR* No 31120, 05/06/2015), see also cases under (29) and (31) and in Konstantinidou (2004: 352 and 2005: 267). Note also the lexicalized case of SMG *ksana mana* again ECHO [+negative] as well as the SMG *ares mares* 'incomprehensible speech, nonsense' and *i sara ce i mara* 'group of people of all kinds [+negative]', considered in this study as TRCV cases containing *mühleme* (in the same direction also Joseph 1985: 92–93), yet made up of two non-words (see 2.2).[15] In these terms, the pattern is marked in SMG as colloquial and informal.

Nevertheless, given the fact that a general treatment of SMG *mühleme* is so far missing, remarks or conclusions regarding the frequency of the pattern in SMG should be considered rather tentative. So far, scholars seem to be in doubt whether this pattern is in common use throughout Greece. Sarantakos (2013) suspects that [m]-reduplication may be a rather characteristic phenomenon in Northern Greece, because it is not very familiar in Southern Greece. Note that [m-] is treated as a lexical entry not only in DSMG (edited 1998) in Thessaloniki (Northern Greece), but also in the very recent UMGD (edited 2014) in Athens (yet with some rather lexicalized cases of *mühleme* as examples of usage, such as *ksana mana* 'again ECHO'). It is also noteworthy that Thomopoulos (1986: 735), the first scholar known to us to observe the phenomenon in SMG, complains at the lack of relevant information in dictionaries, encyclopedias or other textbooks. All this suggests that the presence of *mühleme* in SMG is not by any means uncontroversial, and that the whole issue deserves further investigation (see also Konstantinidou 2004: 353).

---

15 Note that most etymological suggestions in SMG dictionaries tend to consider the base words of these constructions as derivatives from ordinary syntactic words, though facing difficulty to adequately explain the presence of the [m]-reduplicand term (see also fn. 8).

With regard to local varieties of Modern Greek, the *mühleme*-pattern is fully productive in eastern dialects of MG,[16] especially in dialects of Asia Minor, such as the dialect of Pontus (Valavanis 1892, Psaltes 1915, Athanasiadis 1977, e.g., *jineka mineka* woman ECHO, *aspalixton maspalixton* closed ECHO), the Cappadocian dialect (Anastasiadis 1976, e.g., *kotšia motšia* wheat.PL ECHO) or the dialect of Bithynia (Konstantinidou 2004, 2005, e.g., *θies mies* aunt.PL ECHO). Note that the pattern is also attested in older sources of literary dialectal speech, particularly, in two comedies of the 19th century, both written by Constantinopolitans (Byzantios and Alexandros Rizos Rangavis).[17]

The *mühleme* with ordinary lexemes has also been traced in almost all western or island dialects of MG.[18] Yet, it must be emphasized that the majority of the *mühleme* data attested in these dialects has not been detected in everyday discourse, but in somehow established forms of speech, e.g., in folk songs, nursery rhymes, incantations and the like: e.g., *kalanda malanda* 'carol.PL ECHO' (Northern Greece: Halkidiki), *saʎagas maʎagas* 'snail ECHO' (Northeastern Greece, Thrace), *tiri miri* 'cheese ECHO' (Northwestern Greece, Epirus), *sinerya minerya* 'utensil.PL ECHO' (Central Greece, Boeotia), *saligare maligare* based on *saligaros* 'snail' (Southern Greece, Peloponnese), *tsigra migra* 'eye.gum ECHO' (Aegean, Crete), *siliɣuðja miliɣuðja* based on *siliɣuði* 'a type of seprent, lizard' (Eastern Aegean, Samos), *zila mila* based on *ziʎa* 'jealousy, envy' (Cypriotic Greek), *stafiliti mafiliti* based on *stafilitis* 'uvulitis, a disease of the throat' (Cypriotic Greek) (Archive ILNE, Konstantinidou 2004, 2005, Passalis 2012: 12), *kasiði ce masiði* based on *kasiðis* 'scurfy person' (unspecified area, Lelekos 1888: 142). The question whether the *mühleme* represents a living pattern in all these dialectal areas, or whether it is a rather obsolete or restricted pattern, cannot be answered without targeted research. The only example of use in everyday speech so far attested in western dialects of MG comes from Central Greece, *Andzeles ce Mandzeles* Angela.PL and ECHO, based on the name Angela (Western Greece: Aitoloakarnania; datum reported by a commentator in Sarantakos 2013). From Central Greece, note also the instance *kanis manis* 'none ECHO', encountered in a document dated 1815 (Loukopoulos 1938: 447).

16 For the taxononic proposal *eastern* vs. *western* MG dialects see Triandafillidis (1938: 66–68).

17 See the numerous cases of *mühleme* used by a Greek from Asia Minor in the popular comedy Babylonia (1836) of Byzantios (e.g. *piperi miperi* pepper ECHO, Levy 1980; Konstantinidou 2004, 2005), as well as the mühleme case *kurti ce murti* based on *kurti* 'court, flirting' in the comedy of Alexandros Rizos Rangavis *The Fiance of Archondula* (verse 787, in the edition of 1843).

18 There is so far no *mühleme* evidence available from the Modern Greek varieties "Grico" and "Grecanico" in Southern Italy and the Tsakonian dialect (a living descendant of Laconian, spoken in Tsakonia, a group of villages in Eastern Peloponnese).

Certainly, the geo-linguistic distribution and frequency of MG *mühleme*, both as fixed and as productive pattern, deserves further contemplation.

2.3.1.5 Semantics of Modern Greek *mühleme*

Regarding the semantics of Modern Greek *mühleme*-constructions applied to ordinary syntactic words, it is necessary to distinguish between their semantico-functional roles in ordinary/everyday speech on the one hand and their role in specific linguistic registers on the other hand. The fact that *mühleme*-patterns are very often encountered in the registers of verbal charms (incantations) and nursery rhymes (Konstantinidou 2004: 350 and Konstantinidou 2005: 263–264), that is in traditional systems carrying archaic/diachronic elements, provides justification for this distinction.

2.3.1.5.1 *Mühleme* meanings/functions in linguistic registers

The m-reduplicative pattern has been shown to be part of the special linguistic repertoire of Modern Greek charms (along with nonsense words, *artificial constructed compounds* etc., see Passalis 2012). The formation of 'pseudo-words' – as Passalis calls the [m]X components of the m-reduplicative patterns – on the basis of the base words works in the framework of verbal magic as a tool "that aims at handling and controlling the targeted recipient as well as confirming the power of the performer over him so as to force him into obeying his desire" (Passalis 2012: 15). Note that the recipient is usually a disease, a harmful or pesky animal/reptilian, a human disease or similar, e.g., (33)–(34). (See further examples and discussion in 2.3.1.6: Note on origin of the *mühleme*). This implies that in magic speech the use of *mühleme* has a distinct performative ('speech-act') character.

(33) dialectal, Cypriotic Greek [Archive ILNE, s.v. *μαβιλίτης*]

Stafiliti	*maviliti*	*tš'*	*aðerfe*	*tu*	*Haronda*
uvulitis.VOC	ECHO	und	brother.VOC	of	Charon.GEN

'You bloody/damed uvulitis[19] and brother of the Death'

(34) dialectal, Crete [Konstantinidou 2004: 349]

na	*jitepsi*	*ti*	*dzigra,*	*ti*	*migra* [...]	*ce*	*ola*
PTCL	charm.3SG	the	eye.gum	the	ECHO [...]	und	all.PL

19 A disease of the throat (Passalis 2012: 12).

ta *kaka*
the malice/evil
'to charm the eye gum[20] and the like […] and on every evil'

In MG nursery rhymes, which often have a magical substratum,[21] the nature of the pattern is playful but also mocking – the latter role resulting doubtlessly from prior magic uses of the songs, (examples 35–36, as well as 40 below). A ludic, playful function is also attested in tongue twisters, (e.g., 37).

(35) dialectal, unspecified area [Lelekos 1888: 142]
Kasiði *ce* *masiði,* | *poso* *to* *pulas*
scurf.MASC.VOC and ECHO, | how.much CLIT sell.2SG.PRS
to *ksiði?*
the vinegar
'Scurfy and schmurfy, | how much do you sell the vinegar?'

(36) dialectal, Peloponnese: Arcadia [Dr. Christos Dalkos, p.c.]
Saligare *maligare* | *vɣal'* *ta* *cerata* *su* |
snail.MASC.VOC ECHO | stretch.out.2SG.IMP the.PL horn.PL yours
na *pame* *sti* *ɟira* *su* […]
PTCL go.1SG.PL in.the mistress yours […]
'Snail schmail, stretch your horns out, so that we can go to your mistress […]'

(37) dialectal, unspecified area [cf. Konstantinidou 2004: 350]
erçete *o* *kotsifas,* *o* *motsifas*
come.3SG.PRS the blackbird the ECHO
me *ta* *kotsif-* *o-* *motsif-o-* *peðopula* *tu*[22]
with the blackbird-(o)-ECHO-(o)-chick.PL his
'There comes the blackbird the schmackbird with its blackbird-schmack-bird-chicks'

2.3.1.5.2 *Mühleme* meanings/functions in ordinary speech

Firstly, by using a *mühleme*-construction the speaker expresses dislike, disturbance, disapproval, and/or a desire to get rid of the referent or the event denoted

20 A disease of the eyes.
21 For the link that connects nursery rhymes with earlier ritual worship songs or earlier charms see the literature in Passalis (2012: 17 n. 9), also in Konstantinidou (2004: 350 and 2005: 263–264).
22 At https://el.wiktionary.org/wiki/σκουληκομερμηγκότρυπα (checked 05/01/2017).

by the base word X, insofar as this is being experienced as something troublesome or pesky, (see examples 38–39, DSMG s.v. *μ*-, and Konstantinidou 2004: 347, 348). Note that ludic style is not precluded from this use.

(38) SMG [DSMG, s.v. *μ*-]

Komena	*ta*	*ekso*	*mekso*	*ðen*	*procite*	*na*	*ksanavjis*	*ekso*
cut.PTCPL.PRF	the	out	ECHO	not	be.going	PTCL	again.go.out.2SG	out

'An end to the exits and so, you will not go out again!'

(39) dialectal, Bithynia Asia Minor [Konstantinidou 2004: 347]

-Ade	*vre*	*tsifuti!*
INTERJ	INTERJ	stingy.MASC

'-Come on you stingy person!'

-Tsifuti,	*mafuti,*	*ast'*	*afta*	*tora!*
stingy.MASC	ECHO	leave.IMP	these	now

'-Stingy schmingy, cut the crap now!'

Secondly, the use of *mühleme* connotes contempt towards or disparagement of the referent (typically a person). In this case, the pattern serves as a verbal means to mitigate, offend, and ridicule X, (see also examples 29–31, 40 and Konstantinidou 2004: 347, 348). Ludic style is also apparent in this use.

(40) dialectal, Bithynia Asia Minor [Konstantinidou 2004: 347]

kanenas	*ðen*	*irθe*	*sto*	*nosokomio*	*(na*	*ti*	*ði),*
nobody	not	come.3SG.PST	to.the	hospital	(PTCL	her	see)
θanasis,	*manasis,*	*kanenas!*					
Thanasis	ECHO	nobody					

'Nobody came to the hospital to see her. Thanasis Schmanasis, nobody!'

Thirdly, the notion of indifference towards what is referred to by the base may be also attested in some *mühleme*-constructions in MG (Konstantinidou 2004: 348). This notion could be considered as a mild type of contempt or disparagement of X in all possible variations of the construction (see, e.g., 41).

(41) Northern Greece: Thessaloniki [H. Kallergi, personal archive]

-Pu	*ine*	*to*	*Bazaar?*
where	is	the	Bazaar?

'-Where is Bazaar?'

-Bazaar,	*Mazaar,*	*eɣo*	*ta*	*tsiɣara*	*mu*	*mono*
Bazzar	ECHO	I	the	cigarettes	mine	only
ksero	*pu*	*ine*				
know	where	are				

'-(I don't know/care about) Bazaar and such stuff. I only know where my cigarettes are'

Fourthly, a list-like description, i.e., the meaning of 'X and the like', 'X and stuff', and therefore 'X and whatever else relevant', is attested in SMG (Thomopoulos 1986: 735f.)[23], as shown in examples (42)–(43), as well as in dialectal MG (Konstantinidou 2004: 352 and 2005: 261), as shown in examples (44)–(45). This meaning is known from Turkic and many other languages (Müller 2004, Southern 2005 and citations therein), see, e.g., the Aromunian example in Stolz (2008: 117). A derogatory or ludic mood should not be precluded from this use too.

(42) SMG [Thomopoulos 1986: 735]

Tora ti meɣalovðomaða vapsame portes – mortes
now the Holy.Week paint.1PL.PST door.PL ECHO
'During the Holy Week we painted doors and windows and other frames'

(43) SMG [Thomopoulos 1986: 735]

Ferane sto xorjo nero, ce mas evale
bring.3PL.PST in.the village water and us install.3SG.PST
sto spiti o iðravlikos solines – molines,
in.the house the plumber pipe.PL ECHO
tu kozmu ta lefta
of.the people the money
'They have brought water to the village, and the plumber has installed pipes and the like in the house, (so we have spent) a pile of money'

(44) dialectal, Cappadocia [Anastasiadis 1976: 289]

Kotšia motšia tšip pirin ta
wheat.PL ECHO everything take.3SG.PST them
'Wheat and stuff, he took everything'

(45) dialectal, Asia Minor [Byzantios' *Babylonia* A 6, Evangelatos 2002]

Emena fcase me pasturma me t' avɣa [...]
me prepare.SG.IMP CLIT salted.meat with the egg.PL [...]
piperi miperi
pepper ECHO
'For me get ready salted meat with eggs [...] pepper and whatever else relevant'

---

23 According Thomopoulos (1986: 735) the m[X] component denotes in summary synonyms and related names of the base word.

As a fifth function, we should mention the semantic aspect of intensity or emphasis, generally associated with repetition patterns (see, e.g., Kallergi 2015). This is attested in the Pontic dialect (example 46, see also the Kurdish example *xint-mint* 'absolutely crazy' in Stolz 2008: 118, Map 1). A negatively marked, affective connotation (e.g., plaintiveness) may be also included in this use. It is an issue for further research if epistemic modality is also involved, with the pattern functioning as a grammatical marker connoting 'indeed', i.e., expressing strong assertion.

(46) Pontic dialect [Athanasiadis 1977: 32]

Ezisan	*axara*	*maxara*
live.3PL.PST	gracelessly	ECHO

'They lived in great difficulty/with a lot of suffering'

Another important functional dimension of the *mühleme* is that it invites concessive readings of the type *although p > q*, especially when applied to verbs, as in example (47) (Kallergi 2013: 319). Notions such as contempt, emphasis, "free choice quantification" (i.e., 'X and the like'), epistemic necessity, etc. could possibly give rise to concessivity in this and similar types of MG reduplicative patterns (for a discussion, see Kallergi 2013).

(47) dialectal, Northern Greece [Kallergi 2013: 319]

griŋazes	*miŋazes,*	*to*	*fajes*	*olo*	*sto*	*telos*
nag.2SG.IPFV.PST	ECHO	it	eat.2SG.IPFV.PST	all	in.the	end

'you were nagging and all, but you ate it all in the end'

Finally, the *mühleme* in MG may express rejection without having concessive function (this usage is similar to the Yiddish *schm-* reduplication), cf. Zwicky & Pullum (1987: 338). For example:

(48) dialectal, Bithynia [Konstantinidou 2004: 348]

-Pare	*(verikoka)!*	*Ine*	*bebeko!*
take.IMP	(apricot.PL)	be.3PL	bebeco

'-Here, take some apricots! They are bebeco!'

-Bebeko,	*memeko!*
bebeco	ECHO

'Bebeco schmebeco!'

(49) dialectal (in a verse comedy dated 1843) [Rangavis 1874: 415]

Kurti	*ce*	*murti,*	*ci*	*opos*	*ci*	*an*	*lejete,*
flirt	and	ECHO	and	whatever	and	if	be.called.3SG.PRS
ðen	*me*	*aresi* [...]					
not	me	like.3SG.PRS					

'Flirt or whichever may be its name, I don't like it [...]'

Generally speaking, Modern Greek X [m]X patterns may be considered as verbal (grammatical) means used by the speaker in order to express (connote, not denote)[24] predominantly negativity, i.e., negative feelings or negative moods/ attitudes to someone/something (e.g., dislike, disapproval, contempt, indifference). Additionally, a non-affective performative component may also be assigned to the semantics of MG *mühleme*, in the sense that the speaker makes use of such patterns in order to 'harm' (distort) in virtual terms, i.e., to have a negative effect on some pesky or annoying aspect of reality so that he gets rid of it (Konstantinidou 2004: 348), see especially examples (39), (33), (34), (48) and (49). As has been noted in Konstantinidou (2004), Stolz (2008) and Kallergi (2013), the functional effects of the formal distortion seem to iconically reflect the real-life concomitants of actual damage and/or verbal insult. To the performative component one may be ascribe the ludic/comedic aspects of meaning attested in several *mühleme*-constructions, e.g., (39), (45), and (48). For the possibility of a causal relationship between *mühleme* functions in ordinary speech and in magic speech, see below.

For the semantics of fixed *mühleme* expressions when used in everyday, predominantly dialectal speech (e.g., notions such as 'nonsense' or 'muddle, disorder, mishmash'; or '(vaguely) things, bric-a-brac' also known from similar patterns in German[25] and other languages[26]) as well as for the nonsemantics of the fixed *mühleme* when encountered in genres like riddles, nursery or counting-out rhymes and the like related to verbal magic anyhow[27] (see, e.g., the cryptic/puzzling function of totaly unintelligible *mühleme*-constructions in riddles) see section 2.2 of this study.

2.3.1.6 Note on the origin of MG *mühleme*

The oldest attestation of an X [m]X expression in Greek concerns two magical formulas of horse medicine, attributed to Apsyrtus (4th cent.) in *Hippiatrica* (Paris manuscript, where formulas are dated before 9th century):

---

24 For the distinction between (a) cognitive aspects of meaning, i.e., notions related conventionally to linguistic elements and being *denoted* by them, and (b) non-cognitive aspects of meaning, among them feelings or moods related conventionally to linguistic elements and being *connoted* by them see in detail the discussion in Konstantinidou (1997).

25 "Interessant ist, daß praktisch bei allen deutschen Beispielen (Schmorlemorle, Techtelmechtel usw.) ein Mit- oder Durcheinander eine Rolle spielt" (Müller 2004: fn. 164).

26 See literature in Stolz (2008: 119).

27 See Konstantinidou (2004: 350); and citations therein.

(50) Koine/Medieval Greek
[*Hippiatrica Parisina*, *Corpus Hippiatricorum Graecorum* 2, Section 20, line 2]

Πρὸς	*πᾶσαν*	*μᾶλιν.*	*Μὴ*	*ῥήξῃς*	*τὸ*	*κτῆνος*	*τοῦτο,*
toward	every	farcy	NEG.PTCL	rip.2SG	the	beast	this
σαρρα,	*μάρρα,*	*καμέτριξ*					
[non-word]	ECHO	[non-word]					

'Against any kind of farcy[28]. Do not rip apart this animal, *sarra, marra, kametrix*'

(51) Koine/Medieval Greek
[*Hippiatrica Parisina*, *Corpus Hippiatricorum Graecorum* 2, Section 21 line 4]

[...] σαρραμαρρα	*καμετριξ*	*γράφων*	*ἐν*	*χάρτῃ*
[non-word].ECHO	[non-word]	write.PTCPL	in	paper.DAT
περίαπτε	*εἰς τὸ κάπιστρον*			
pass.around.2SG.IMP	in the halter			

'[...] after writing on a paper *sarramarra kametrix*, pass it around the halter'

These terms are nothing else but *voces magicae, omomata barbara, nomina barbara* or abracadabra-words, i.e., lexically nonsemantic words,[29] encountered usually in magical papyri, incantations, invocations, conjurations and the like.[30] According to Kopidakis (2007), the SMG phrase *i sara ce i mara* 'group of people of all kinds [+negative], mishmash', for which there are several etymological proposals – especially in respect to the term *mara-* but none of them

28 A kind of glanders.

29 According to Versnel (2002: 108) a word of the type *abracadabra* is an expression "which certainly *has* a sense (that is: a function, an objective), but which does not *make* sense: it does not carry a comprehensible, lexical meaning".

30 Eusebius, Bishop of Caesaria (PG Migne 10.10.9) considers these terms as "chamber's tricks and cover-ups" (*«γοήτων [,,,] ταῦτα πάντα τεχνάσματα καὶ προκαλύμματα»*). According to Krulak (2016: 82), "the *onomata barbara* did have meaning and was a legitimate source for its communication; the meaning was inaccessible to humans, but this was not the intended audience. The gods understood the meanings behind the invocations, responded to them [...]". Similarly Mauss & Hubert ([1902] 2002: 35–36, after Passalis 2012: 8): "Les incantations sont faites dans un langage spécial qui est le langage des dieux, des esprits, de la magie" and Malinowski (1922: 432, in Passalis 2012: 8): "a considerable proportion of the words found in magic do not belong to ordinary speech, but are archaisms, mythical names and strange compounds, formed according to unusual linguistic rules". For a discussion on the issue and relevant literature see Versnel (2002: 114ff., also fn. 26, 33), Kopidakis (2007: 251–252), Passalis (2012: 9, fn. 3), Eideneier, *Spanos* (1977: commentary in pp. 200, 236, fn. 60).

quite satisfactory,[31] should be derived from the above mentioned Koine or Medieval Greek *voces magicae*.

Notably, an instance of m-reduplication, applied also to a *vox magica*, could also be indentified in a Latin popular spell against tooth-ache attested in *Marcellus Empiricus*, a medical writer from Gaul at the end of the 4th and beginning of the 5th century:

(52) Latin [Marcelli de medicamentis, *Corpus Medicorum Latinorum* 5, XII, 24]
argidam margidam stvrgidam

For the assignment of these words to *voces magicae* see Versnel (2002: 131, 133); for an attempt to make sense of this spell "by means of etymological engineering"[32] see Knobloch (1989).

It is important to mention in this context that it is a very characteristic rhetorical strategy of the language of the magical formula for achieving its goals to "repeat a *vox magica* once or twice, either without or with slight alterations, after which there is a sudden radical change in the following element(s) ("*argidam margidam sturgidam*")" (Versnel 2002: 155).

Evidence related to m-reduplication of lexically nonsemantic words is to be found also in Medieval Greek. The relevant expression, mentioned already under (12) and repeated here for convenience, appears in a parody of a magical-medical formula in *Spanos*, an anonymous vernacular text of the 14th–15th cent., within which various texts with ecclesiastical, law and magical-medical content are satirized.

(53) Medieval Greek
[Eideneier, *Spanos* 1977, B 49-50, MS -*ντσ*-, Kriaras, *Epitome* s.v. *άταλα* 2]

(μάζωξε)	*ἄντζαλα*	*μάντζαλα*	*σάντζαλα*
collect.IMP	[non-word].N.PL	ECHO	ECHO

'(collect) andzala mandzala sandzala' (i.e., 'unimportant or nonexistent things', Kriaras, *Epitome* s.v. *άταλα* 2)

The word *andzala* (with the variations *antala, antsala*), explained by some scholars with the notion *nugas* (Latin) 'trash, trashy/worthless things that are to be thrown away' (see commentary of M. Crusius in Eideneier, *Spanos* 1977: 243), has been characterized as artificial, and as a facetious neologism. This also applies to the words *mandzala* (also *mandala*) and *sandzala* (also *sandala*) (Eideneier, *Spanos* 1977: 266–267, Eideneier, *Spanos* 1990: 105; see also Kriaras,

31 Beside Kopidakis (2007) see also Papadopoulos (1953: 16–18).
32 See Versnel (2002: 131, fn. 68).

Epitome s.v. ἄταλα 2, Kriaras 1968, s.v. μάνταλα, σάνταλα). The anonymous writer of *Spanos* exploits here the scaffold of well-known nonsense magical formulas, yet not for a serious purpose, but in order to spoof, insult or mock jokingly (see also Eideneier, *Spanos* 1977: 199) – a sign of the decline, debunking or derision of magic. (Note also *andzala mandala sandala* a variation of this expression attested in *Spanos* A 225 ed. Eideneier 1977: 111).

These diachronic data, all encountered in magical-medical formulas and sharing a recognizable *mühleme*-like form and a cryptic, unintelligible, confused or scrambled, at least for the layman, meaning along with later joking, scornful, contemptuous connotations, suggest a direct connection between MG frozen (lexicalized) *mühleme*-constructions of an obscure origin (see cases treated in 2.2) and nonsense magical formulas (see type identified in Versnel 2002: 155). By deriving the former (i.e., *mühleme* of an obscure origin) from the latter (i.e., lexically nonsemantic words of magic), basic semantico-functional aspects of these fixed m-reduplicative constructions (such as the notions of 'nonsense' or 'muddle, disorder, mishmash' or 'different undefined/unimportant things' along with their [+negative] marks/connotations), can be easily justified. It is assumed that in a time of demystification/derogation of magic *voces magicae* (i.e., meaningless, incomprehensible, confused and therefore worthless/rubbishy for the laymen words of magic speech, just like the words of an unknown language) are taken out of their magical context (or become the subject of imitation) and used for non-magical purposes, or better misused by laymen as metaphors in order to characterize the referent in negative/derogatory terms, e.g., a set of incomprehensible words, a host of unimportant/trushy things, an inhomogeneous group of persons. In addition, by relying on the techniques of the language of magic, the cryptic function of the fixed *mühleme*-patterns in riddles, as well as their frequent occurrence in these and similar linguistic registers can be also easily justified. Besides this, the etymological obscurity of these patterns could find a plausible reading.

Here, it may be (tentatively) suggested that whenever a fixed TRCV pattern that can be analyzed in terms of a *mühleme* is found bearing meanings such as 'nonsense, disorder, confusion' and the like, no etymological attempt to connect it with ordinary lexemes should be undertaken, insofar as it will doubtlessly be a non-word case, modeled after the modi of the nonsensical vocabulary of verbal magic (see under this perspective, e.g., some relevant cases in 2.2). This recommendation may also apply to other etymologically obscure fixed expressions with a recognizable m-reduplicative structure, meaning (rather vaguely) '(different undefined/unimportant) things, trifles' and the like, e.g., *dzadzala madzala, dzidzili midzili, dzidzi midzi* etc. A case in point is in our opinion also the etymologicaly opaque SMG expression *ta sea ce ta mea mu/su/tu* (*τα σέα και τα μέα μου/σου/του*) meaning vaguely 'things, whatever belongs to or is characteristic of me/you/him' without specification, analysis or exact naming of the referent. Expressions of this kind may find a satisfactory reading if treated like the *mühleme*-patterns encountered in riddles, i.e., like non-words that are naming (vaguely) whatever cannot be named by its ordinary/conventional name, due to the cryptic purposes of the riddles.[33] Note that a connection of some SMG frozen phraseological pairs such as *ares mares* 'nonsense, gibberish' or *i sara ce i mara* '(group of) people of all kinds [+negative], trash in metaphorical terms' with the pattern X [m]X has been already established by scholars like Joseph (1985) and Kopidakis (2007).

As regards the origin of the MG productive *mühleme*-pattern applied to ordinary words, we may also suggest that it derives from magic, cryptic language (Konstantinidou 2004, 2005). Two arguments point to this direction: a) the connection claimed above of fixed *mühleme* with nonsense magic formulas, and b) the identification of the productive *mühleme*-pattern as a specific means in MG verbal charms for the alleged modification of extra-textual reality (Passalis 2012).

Basing *mühleme* as a productive pattern on verbal magic justifies the performative ('speech-act') functions of the pattern, i.e., the notion of 'harming' somebody/something by words, as in magic, as well as the negative affectivity (contempt, annoyance, indifference and the like) addressed to the annoying or troublesome referent that the speaker wants to get rid of, as the referent in magic. It can also justify the playful, ludic aspects of the pattern in genres with a faded magical substratum like nursery rhymes (as in example 35 above) tongue-twisters and the like. The notion 'and the like' may have easily resulted from the mere formal/morphological similarity of the m[X] component with the base word, when considered from the perspective of a layman, i.e., when taken out of its original magical context. It is no coincidence that a prefix [m-] cannot be easily identified as a grammatical element (morpheme) bearing a distinct kind of lexical meaning in (Modern) Greek,[34] hence the absence of the pattern from Standard and dialectal MG grammatical descriptions, with very few exceptions until now (mentioned above, see 2.3.1). Incidentally, if [m] actually derives from the field of magic, one might wonder whether the choice of this particular

33 Cf. the cryptic function of SMG *ameletita* 'lamb's fries' vs. *orçis* 'testicles, usually of slaughtered animal', due to shame or taboo.

34 This also applies to Turkish and other languages where *mühleme* is also attested (Müller 2004: 326).

sound is accidental "or do we really have to do with a kind of phonemic symbolism" (Passalis 2012: 15).

In spite of the above, we must admit that the MG productive *mühleme* mechanism may also have been reinforced by language contact in the domain of the former Ottoman Empire (Stolz 2008). The pronounced presence of the MG pattern in Eastern MG dialects (e.g., Pontic, Cappadocian, Bithynian), which have coexisted for centuries with Turkish, argues for this interpretation. Yet, the semantico-functional diversity and the overall complexity[35] of the MG pattern (especially its performative aspect, its applicability to every word-class, its productivity), when taken into account along with other parameters, such as the geo-linguistic distribution of the pattern in many typologically diverse languages (Müller 2004, Southern 2005, Stolz 2008), point to the necessity for further research in the direction suggested in this study. All in all, careful research in the domain of magic material (oldest and newest)[36] could possibly shed some light on the origins of this reduplicative phenomenon. If the relation between *mühleme* and magic speech could be confirmed, the idea of the universality of the pattern would certainly come to the fore.

2.3.2 Reduplication with *kse*-prefixation/*X kse-X* constructions

Among all the Greek productive constructions involving reduplication with fixed morphemes (prefixes), the pattern *X kse-X* stands out on the basis of its high productivity and frequency.

In more detail, the pattern is created as follows: a word of any major category (mostly, though, nouns and adjectives) and, theoretically, of any case type (except for the vocative) may be juxtaposed to its copy bearing the prefix *kse-*, as in the following examples (54)–(58). These express indifference (54), rejection (55), concession (56), the "X and the like" meaning (57) and intensity/emphasis (58) respectively:

(54)	*Ti*	*na*	*kano*	*re*	*Mits-ara,*	*afu*	*stin*	*Turcia*
	what	to	do	INTERJ	Mitsos-AUG	since	in.the	Turkey

35 According Müller (2004: 326) the complexity degree of the *mühleme*-pattern in Abkhaz calls into question the assumed derivation of the pattern from the Turkish one.

36 See Borsje (2011) for the pilot project of the University of Amsterdam and the Meertens Institute under the supervision of Jacqueline Borsje of compiling a digitized database with charms of different cultures and eras.

papas-kse-papás	*ðen*	*ta*	*ɣlitonis*	*ta*	*tria*
priest *kse*-priest	not	them	escape.2SG	the	three
xroɲa	*fadarilici*[37]				
years	military.service				

'What should I do, Mitso (pal), since in Turkey whether you are a priest or not, you can't escape the three years of military service'

(55)

(Ti) papas	*ce*	*kse-papás*	*oli*	*ta*	*iðja*	*ine*[38]
what priest	and	*kse*-priest	all	the	same	are

'It doesn't matter (/I don't care) if he's a priest (/What nonsense are you talking about when saying he's a priest); they are all the same'

(56)

[...]	*krio*	*kse-krio*	*θa*	*pao*	*sto*	*londino* [...][39]
	cold	*kse*-cold	FUT	go.1SG	to-the	London

'Despite the cold, I'll go to London [...]'

(57) [Anastasiadis 1976: 289ff.]

sitaria	*kse-sitaria*	*ola*	*ta*	*pire*
wheat.PL	*kse*-wheat	all.PL	them	take.3SG.PST

'S/he took everything, wheat and all these'

(58) Pontic dialect [Papadopoulos 1958, s.v. *ξέκαλος*]

kalós	*ce*	*ksé-kalos*
good	and	*kse*-good

'very good'

Examples (54), (55) and (56) represent the most frequent functions/meanings for the pattern in question, whereas (57) and (58) are relatively rare cases (with the latter being only dialectic, but see also the analysis below).[40]

Regarding the morpheme *kse-*, this appears as a highly productive prefix in Greek, which mainly applies to verbs as well as deverbal nouns and adjectives

37 http://www.esoterica.gr/Forums/topic.asp?whichpage=13&ARCHIVEVIEW=&TOPIC_ID=4860 (comment by junk95, 16/12/2004).

38 At http://www.remaliaclub.gr/forum/showthread.php?t=2106&langid=1 (last access January 1st, 2017).

39 At http://sfrang2.blogspot.com/2012/02/blog-post_19.html (last access October 5th, 2016).

40 Specifically with respect to example (57), it appears as a translation of the Pontic example *Kotšia motšia tšip pirin ta* 'wheat and stuff, (s)he took everything'. The use of a *X kse-X* pattern seems to have been chosen by the author (Anastasiadis 1976) as an equivalent expression that sounds more familiar to speakers of the standard variety of Greek. Another possible appearance of the meaning 'X and the like' is again dialectic, hence, we cannot be absolute about its interpretation:

(i) (Aegean, Skiathos)

Pu	*loɣarjaza*	*talaro*	*kse-talaro!*	*Naaaa!*	*Fovos!*
That	count.1SG.IPFV	penny	*kse*-penny	INTERJ	fear

'Me, that counted pennies and such! I was so frightened!'

with a wide range of meanings. Some of these meanings are i) contrast, in the sense of removal, cancellation and/or reversal of action (e.g., *dipsao* 'be thirsty' > *ksedipsao* 'stop being thirsty', *dino* 'dress' > *ksedino* 'undress') and ii) absolute intensity (*ksekufeno* 'make somebody completely deaf', *ksetrelenome* 'be utterly excited').[41] As mentioned in Ralli (2005: 42), *kse-* is a Modern Greek prefix, in the sense that it did not exist in Classical Greek, but was created during the Medieval period on the basis of the Ancient Greek preposition *ek-* 'from'.

In the *X kse-X* pattern, specifically as exemplified by (54)–(56), the prefix *kse-* does not seem to perform its typical role (that of creating existing MG lexemes), but rather appears to serve the purpose of creating neologisms or, better, *hapax legomena* (i.e., ephemeral words/expressions that have not entered the lexicon). This observation is based on a number of factors. First, within this pattern *kse-* does not impose its normal word-class restrictions, but, as aforementioned, it can apply to words belonging to any part of speech (including particles, conjunctions and interjections), as well as extend its scope to clauses, as in (59) and (60) respectively:

(59) SMG [DSMG 1998, sv. ξε-]

den	*eci*	*ma*	*ce*	*kse-ma!*
not	have-3SG.PR	but	and	*kse*-but

'there is no "but" and such things!'

(60)

Pame	*me*	*to*	*stavro*	*sto*	*ceri.*	*Mas*	*rixnune*	*kse-mas*
go-1PL	with	the	cross	in-the	hand	us	drop-3PL	*kse*-us

rixnune.	*Den*	*milame!*[42]
drop-3PL	not	talk-1PL

'We stick to morals; whether they do us wrong or not. We don't react!'

Second, in combination with nouns and adjectives, typically *kse-* can cause a change in the stress pattern of the base, as, e.g., in *kséstrotos* 'unlaid' (but *strotós* 'laid even, easy') (see also Ralli 2005: 242–243). By contrast, as is observed in (54)–(55), *kse-* in the *X kse-X* construction does not affect the stress of the output word (*ksepapás* instead of *ksépapas*).

Thirdly, with nominal bases, typical *kse-* obligatorily causes vowel deletion when the nominal base starts with a vowel (Ralli 2005: 243): *aniɣo* 'open'> *ksaniɣo* (*kse-aniɣo*) 'move further away', *amolo* 'let' > *ksamolo* (*kse-amolo*) 'let

41 For other meanings and functions of the prefix *kse-* see Hatzidakis (1914), Efthymiou (2001, 2002, 2003) and Papanastasiou (2012).

42 The instance was attested in Athens (taken from Magdalene Konstantinidou's personal record).

free'. However, with neologisms, vowel deletion is not obligatory (compare examples 62 and 63):

(61) *me* *ton* *cero ton* *kse-agapise*[43]
with the time him *kse*-love.3SG.PFV
'With time, she stopped loving him'

(62) *arrosti* *kse-arrosti,* *spiti* *ðe* *gaɵome*[44]
ill.FEM *kse*-ill.FEM house not stay.1SG.PR
'Ill or not, I'm not staying home'

(63) *arrosti* *ksarrosti* *ta* *ɣusta* *su* *ta* *vɣazis*[45]
ill.FEM *kse*.ill.FEM the preferences yours CLIT get.2SG
'Ill or not, you get what you like (you have a good time)'

Finally, the copy of the reduplicative construction with *kse-* typically does not appear independently in MG. Some exceptions seem to appear in dialectal data, such as (64)–(65) below:

(64) SMG
den *eçi* *θeós* *ce* *kse-θeós!*
not have-3SG.PR God.NOM and *kse*-God.NOM
'There is no God and such things!'

(65) dialectal, Paros [Archive ILNE]
gamo *ton* *kse-θeó* *su!*
fuck.1SG the *kse*-god.ACC yours (swearing)

In (65), *kse-* may express either intensification or contrast/opposition –being, in the latter case, similar to the SMG word *antíθeos* 'anti-god' "devil" (in the equivalent swearing expression *gamo ton antíθeó su*, which is somehow more wide-

43 This instance was attested in personal oral conversation (Haritini Kallergi, December 29th 2016).

44 https://www.google.gr/search?client=firefox-b&dcr=0&ei=hir-WaWYG9DSwQKCt7O4Cw&=%22%CE%B1%CF%81%CF%81%CF%89%CF%83%CF%84%CE%B7+%CE%BE%CE%B5%CE%B1%CF%81%CF%81%CF%89%CF%83%CF%84%CE%B7+%CF%83%CF%80%CE%B9%CF%84%CE%B9+%CE%B4%CE%B5%CE%BD+%CE%BA%CE%B1%CE%B8%CE%BF%CE%BC%CE%B1%CE%B9%22&oq=%22%CE%B1%CF%81%CF%81%CF%89%CF%83%CF%84%CE%B7+%CE%BE%CE%B5%CE%B1%CF%81%CF%81%CF%89%CF%83%CF%84%CE%B7+%CF%83%CF%80%CE%B9%CF%84%CE%B9+%CE%B4%CE%B5%CE%BD+%CE%BA%CE%B1%CE%B8%CE%BF%CE%BC%CE%B1%CE%B9%22&gs_l=psy-ab.3...1576.6413.0.6997.4.4.0.0.0.0.188.535.0j3.3.0....0...1.1.64.psy-ab..1.0.0....0.SV-rpxGTtAw (last access 24/3/2017).

45 At http://mane-tarious.blogspot.gr/2008/11/blog-post_25.html (last access 24/3/2017).

spread). In fact, the role of *kse-* here seems to be more expressive than semantic (and being expressive, its meaning is less easy to define, see, e.g., Potts 2007). However, more often than not, it is relatively clear whether a normal *kse-* derivative involves the intensifying or the contrastive *kse-*. For instance, in dialects, one may encounter adjectives carrying the *kse-* prefix as an intensifier (example 66), just as it does with verbs (example 67):

(66) dialectal, Crete[46]

mia	*ksé-kali*	*mera*	*tu*	*Mayu*
one	*kse*-good	day	of	May

'On an excellent day of May'

(67) dialectal couplet, Crete[47]

ce	*kse-vareθika*		*pistos*	*ston*	*erota*	*na*	*meno*
and	*kse*-bore.PASS.1SG.PST	loyal	to.the	love	PTCL	stay.1SG	

'and I got totally bored by staying loyal to love'

Also, in some Greek dialects, *kse-* with nouns may signify a no-longer existing quality or characteristic, as in *ksépapas* 'a former priest', *ksedáskalos* 'a former teacher' (Andriotis 1951: 236 reporting from Hatzidakis 1905: 32, see also Archive ILNE, Crete, s.v. *ξέπαπας*). This meaning, according to Andriotis, is taken by analogy from the cancellation/removal meaning that the prefix has with verbs. In SMG, the constrastive/cancelling *kse-* with verbs is also used in neologisms, as shown by example (61) above.

In contrast to these latter cases (examples 66–67), the *kse*-derivatives that appear in the *X kse-X* construction with the aforementioned prevalent meanings (see examples 54–56) cannot be claimed to bear the meaning "very X", "un-X/non-X" or "formerly X"/"ex-X". In fact, the second constituent of this reduplicative construction cannot be said to have meaning in the strict sense. The meaning of *kse-X* in *X kse-X* is only part of the overall meaning of the construction. Taking the semantics of (54)–(56) into consideration, the echo-word acquires either the interpretation "something similar to X" (examples 54 and 55) or "non-X" (in the concessive use, example 56). This is an *ad hoc* interpretation, not a meaning with which the word may appear elsewhere, as an independent lexeme. In sum, although *kse-* normally forms derivatives, within the X *kse*-X pattern it seems to form part of an occasional/incidental derivation, a pseudo-

46 Accessed at https://rethemnosnews.gr/2015/07/το-μεγάλο-τάξιμο-η-γέννα-της-ευγενίας/30th July 2015.
47 Attested in a folklore music TV show (Magdalene Konstantinidou's personal record).

word, which does not exist independently in the language system and cannot be treated as a lexical entry. The prefix itself seems to function similarly to a fixed segment (e.g., similarly to [m] in the echo-word construction), in that it adds a somewhat less clear meaning than usual, but a rich pragmatic/expressive load.[48]

Regarding the possible syntactic manifestations of the construction, the same possibilities with *X [m]X* (the *mühleme*, Section 2.3 above) seem to arise. The constituents of the *X kse-X* pattern may be connected with *ce* 'and' as well as the *(ute...) ute* conjunction ('(neither...n)or...'), as in (68):

(68)	*ute*	*ma,*	*ute*	*ksema,*	*afto*	*ine*	*to*	*programa*	*mexri*
	neither	but	nor	*kse*-but	this	be.3SG.PR	the	program	until
	to 2020 [...][49]								
	the 2020								

'There is not "but" and such (no objections allowed), this is the program until 2020 [...]'

Also, as shown by example (60) above, the constituents of reduplication may be separated by significant syntactic boundaries (that of an independent clause), but the presence of the unstressed clitic *mas* allows reduplication in phonological terms.

One last syntactic feature that should be emphasized regarding *X kse-X* reduplication is the role of the conjunction *ce* 'and'. The presence of *ce* seems to be optional at a first glance, but it may correlate with the actual interpretation of the construction in various contexts. More specifically, example (55) may appear either as in (69) or (70):

48 In the dictionary of Babiniotis (1998), it is claimed that *kse-* has the meaning of removal/cancellation in "occasionally opposite words" ("*επ' ευκαιρία αντιθετικές λέξεις*"), as in *dáskalos ksedáskalos, den eci idea* 'he may be a teacher, but he has no clue'. However, although the meaning of the reduplicative expression *dáskalos ksedáskalos* may be "teacher or not", the word *ksedáskalos* does not seem to independently mean "non-teacher" (as, e.g., *kse-agapo* means "stop loving"), nor does it coincide with the dialectal *ksedáskalos* "ex-teacher". In DSMG (s.v. *ξε*-V), the meaning of the reduplicative pattern *X kse-X* is also considered to be taken by analogy from antonymic pairs of verbs such as *kliðono* – *ksekliðono* 'lock' – 'unlock'. In general, although we agree that in the *X kse-X* pattern under discussion *kse-* has more probably the function of creating opposites rather than the function of intensifying (see, also Kallergi 2013), we suggest that *kse-X* means "non-X" only within the reduplicative construction, hence it does not carry meaning in the typical sense.

49 At http://www.capital.gr/forum/thread/2559737 (last access December 29th, 2016).

(69) *Ti papás ce ksepapás, oli ta iðja ine.*
what priest and *kse*-priest all the same are
'What nonsense are you talking about (when saying he's a priest). They are all the same anyway.'

Here, the presence of *ti* 'what' seems to make the presence of *ce* necessary. Without *ti*, the construction is also possible with *ce*, but then a different intonation would be needed in order to make *X kse-X* sound like an independent clause or an interjection:

(70) *Papás ce ksepapás! Oli ta iðja ine.*
priest and *kse*-priest all the same are
'[You say he's a priest]. That's nonsense! They are all the same anyway.'

The construction in (70) is strongly quotative (it definitely echoes a word preceding in discourse) and serves as a denial of what another interlocutor has said (see, e.g., Kallergi 2013).

In contrast to the aforementioned uses, *X kse-X* does not allow the presence of *ce* when it functions as a concessive clause, as in (56) above. This observation seems to be supported by a large number of instances found in everyday discourse (particularly on the web or the media), such as (71):

(71) *-pços ine o meyaliteros fovos sas?*
which is the biggest fear yours
'What is your greatest fear?'
*-i monaxça mu. Klisé-kseklisé/*Klisé ce kseklisé,*
the loneliness mine cliché *kse*-cliché
aftos ine [...][50]
that.MASC is
'My loneliness. It may be a cliché, but that's it [...]'

In examples like (71), where no quotation is at play, the insertion of *ce* would not be suitable or, in fact, acceptable.

By contrast, the emphatic/intensifying function seems to require the presence of *ce*.[51] As aforementioned, this function is particularly evident in dialects

50 Taken from an interview with an actor in *Τηλεθεατής* magazine, v. 1442: 7

51 There are two instances found among our data that have an emphatic function but do not involve *ce*:

(ii) *anascela-ksanascela* [Pontic, Papadopoulos 1958]
supinely *kse*-supinely
'completely lying on one's back'

such as the Pontic, whereas in SMG it appears marginally (examples 73 and 74), or it often involves the prefix *para-* instead of *kse-* (as in example 76, which is semantically equivalent to 75 in expressing strong assertion):

(72) Pontic [Papadopoulos 1958, s.v. *ξέκαλος* and *ξέκαλα*]

kalós	*ce*	*ksé-kalos* /	*kalá*	*ce*	*ksé-kala*
good	and	*kse*-good /	well	and	*kse*-well

'very good'/'very well'

(73)

fisao	*ce*	*kse-fisao*[52]
blow.3SG.PR	and	*kse*-blow.3SG.PR

'to breathe heavily due to intense stress, anxiety, discomfort or anger'

(74)

-Ejines	*kala?*
become.2SG.PFV	well

'Did you recover (from illness)?'

-Oçi!	*Vixo*	*ce*	*kse-vixo!*
no	cough.1SG.PR	and	*kse*-cough.1SG.PR

'No! I cough a lot/I have this nasty cough!'[53]

(75) Pontic [Papadopoulos 1958]

ime	*ce*	*ex-ime*
be.1SG.PR	and	*eks*-be.1SG.PR

'I am, and very much indeed.'

(76) SMG

ime	*ce*	*para-ime*
be.1SG.PR	and	*para*-be.1SG.PR

'I am, and very much indeed.'

This construction is different from the *X kse-X* instances in (54)–(56) also in that the copy of the reduplication may be phonologically affected by the prefixation with *kse-*: for instance, the stress in *ksé-kalos* and *ksé-kala* rises to the antepenultimate, as with other normal *kse*-derivatives. The fact that the output is phonologically affected by *kse-* prefixation implies that the second constituent

---

(iii)

arostisa,	*mana'm,*	*ksarostisa*	*sti filaci*	*o kaimenos*
got.sick	mother mine	got.sick	in-the jail	the poor

'I, the poor, got sick, mother, very sick, in jail...' [folk song, dialectic, Archive ILNE]

52 See the Utilitarian Modern Greek Dictionary of the Academy of Athens [Χρηστικό Λεξικό της Νεοελληνικής Γλώσσας, 2014] and Babiniotis 1998 (s.v. *ξεφυσώ*).

53 Personal use by Magdalene Konstantinidou (Athens, 2016).

(bearing *kse*) may stand as an independent word, and, in fact, this is exactly the case with most of the examples (73)–(76): *ksékalos*, *ksefisao*, and probably also *exime* (at least, by analogy to *paraime*) are existing MG lexemes or, at least, possible lexical entries.

Last, but not least, the emphatic/assertive *X ce kse-X* has a different syntactic status than the rest *X kse-X* cases: when the pattern expresses rejection/indifference and concession, it serves as a subordinate (adverbial) clause. In contrast, when it is emphatic/assertive, it forms an independent clause or phrase.

Thus, we are most probably faced with two distinct patterns of reduplication with *kse-* prefixation, which bear differences with respect to formal characteristics and functional potential.

One pattern can be roughly represented as *X (ce) kse-X (,/; Y)* whereby the first part of the relation expresses objection or indifference to something stated in the discourse or to *Y* (the statement that may immediately follow). The constituents of this construction relate between each other in either a synonymic or antonymic way; when the pattern expresses objection/indifference ("I don't care about X and such things"), *kse-X* is presented as similar to X; when it expresses concession ("(whether) X or not, Y"), *kse-X* represents a kind of "non-X". It is also possible to think of *kse*-X in the concessive reduplicative construction as a word that means "X and whatever else relevant", i.e., as a term that expresses semantic similarity with X but which, no matter how similar, cannot affect the overall statement in Y. In other words, the scope of "non-" in the concessive construction can alternatively be considered to affect the whole (synonymic) pair consisting of X and *kse*-X.

The other pattern can be represented as *X ce kse-X* and expresses emphasis/intensity/strong assertion. Here, *kse-X* probably stands in a synonymic relation to X, being "(a kind of) X in a very high degree". However, as we saw above, in this construction one may encounter existing lexemes represented by the *kse*-X constituent (e.g., Pontic *ksékalos*). Thus, it is questionable whether the Pontic pattern can be included in the category of reduplication with variation/distortion. Additional evidence from Pontic would be necessary to confirm this observation (e.g., if the intensive pattern is productive in this dialect and systematically involves the juxtaposition of existing lexemes that differ only in that the second one bears the prefix *kse-*). In SMG, the pattern seems to be marginally productive, but, as shown in example (74), the *kse-* constituent (*kse-vixo* '*kse*-cough') is a non-word, an occasional formation with a less than clear meaning.

The difference in functional potential between the two *kse-* constructions may relate to the fact that *kse-* is a polysemous prefix. Thus, it may be the case that a different meaning of *kse-* (the contrastive or the intensifying one) operates in each case. Alternatively, the polysemy of *kse-* may make it vague enough to be used as a fixed segment that does not define the meaning of *kse-X* in expressive constructions, but simply signals the one or the other construction "with the aid of" formal idiosyncrasies, such as the ones we saw above, plus suprasegmental features (intonation).

Despite their differences, *X kse-X* constructions of both semantic directions seem to operate as phraseological units or fixed expressions (in the specific sense of Fraser 1970, Fleischer 1982: 12–13, Chioti 2010: 7) and, even more specifically, as *ad hoc* fixed expressions of the specific kind of phraseological word pairs (twins) related in a synonymic or antonymic way (e.g., MG *fovos ce tromos* 'fear and terror' "very fierce", *sixna pikna* 'often densely' "quite often", *çimona kaloceri* 'winter summer' "all year long", *lini ce ðeni* 's/he unties and ties' "s/he has great power", see Chioti 2010: 111–112, Fleischer 1982: 22, Setatos 1994). Apart from their meaning or cognitive content, expressions of this kind always have affective connotations as well (Chioti 2010: 126–127, Setatos 1994: 188, see also Konstantinidou 1997).

In the case of reduplication with *kse-*, the connotations/pragmatic implications are typically negative. Even in "neutral-sounding" cases like those having the meaning "X and the like", what is represented by *kse-X* seems to be discarded as irrelevant, annoying, unwanted or even ridiculous. The only cases that seem to have positive instead of negative connotations are a few instances of the intensive/emphatic/assertive pattern (e.g., in *kalós ce ksé-kalos* "very good indeed", intensity seems to be appreciated). However, as aforementioned, such instances are rare and unfamiliar to speakers of SMG (and their status as pertinent cases of reduplication with distortion cannot be yet confirmed).

Generally speaking, the areal distribution of *X kse-X* reduplication cannot be stated with certainty. According to Sarantakos (2013), *X kse-X* reduplication of the first type (*X kse-X (, Y)*) is used more in Southern Greece, while *X [m]X* reduplication is more common in Northern Greece. Nevertheless, the pattern is attested also in Northern Greek regions and local dialects (cf. cases from the Archive ILNE mentioned above), along *X [m]X* reduplication. It is the latter, actually, that seems to be confined in Northern dialects (esp. dialects that have been more closely affected by Turkish, see 2.3.1.4), whereas *X kse-X* seems to be ubiquitous in Greece (see entries in SMG dictionaries, e.g., Babiniotis 1998, DSMG, UDMG, s.v. *ξε-*).

Finally, with respect to the origin of the pattern, a few notes are in order. The oldest *X kse-X* expression (in the form *X ex-X*) is attested in an incantation in the anonymous medieval vernacular text of the 14th–15th century *Spanos*

(*Σπανός*, ed. Eideneier), which parodies certain ecclesiastical texts and satirizes texts with legal and magical-medical content.

> "καὶ εὐχοῦνται καὶ λέγουν• ἄ γ κ α ν ο ς , ἐ ξ ά γ κ α ν ο ς , ἑξῆντα λύκοι ἀπάνω είς τὸ κεφάλι σου" Eideneier, *Spanos* 1977, B 217 (Vat. gr 1134, A.D. 14/15?)
>
> "and they wish and say: aganos, eksaganos, sixty wolves upon your head"
>
> "καὶ εὐχόμενοι ἔλεγον: ἄγ κ α ν ο ς , ἐ ξ ά γ κ α ν ο ς | ὁ ποταμὸς ὁ πύρινος | ὁ κατακεκαυμένος" Eideneier, *Spanos* 1977, D 1776 (ex editione Veneta a. 1553)
>
> "and they wished and said: aganos, eksaganos, the river of fire, the burnt-down one"

Kriaras (1968) considers the medieval *eksaganos* (*εξάγκανος*) as an artificial/mock word (a pseudo-word) and assumes it derives from the preposition *ἐκ* (*ek*) and the substantive *aganos*.[54] Note that Eideneier *Spanos* (1990: 105–106) considers *eksaganos* (as well as *aganos*) as an artificial jocular word belonging to the typology of magic. Taking the aforementioned semantic analysis of *X kse-X* patterns into account, we can assume that the phrase *aganos, eksaganos* in *Spanos* represents a case of synonymic phraseological pair, used for the emphatic expression of negativity towards the referent (presumably *Spanos* which means the "beardless or hairless man").

The same pattern can be also found in the title of *Spanos* A 1 (*Ἀκολουθία … σπανοῦ τοῦ οὐρίου καὶ ἐξουρίου* "service…of the beardless man, the *urios*.SG.GEN and *eksurios*.SG.GEN") as well as in a parody of an ecclesiastical ode in the same text (Spanos D 712 *«οὐκ ἐλάτρευσεν/ὁ οὔριός τε καὶ ἐξούριος/τὴν γενειάδαν αὐτοῦ»* the *urios* and *eksurios* did not worship his beard"). The base word *urios*, which in other medieval texts means 'blasted, depraved' (Kriaras 1968, s.v. *οὔριος*) is semantically strengthened in *Spanos* via the prefix *eks-* (as a predecessor of MG *kse-*), and the new derivative word *eksurios* co-occurs with the base word, generating an emphatic phraseological pair. Note that according to Kriaras (1968) both *eksurios* (not attested in other sources apart from *Spanos*) and *urios* (as it is attested in *Spanos*) are merely characterized as abusive words, without any further determination of their meaning.[55] It is also noteworthy that

54 See Kriaras, *Epitome* s.v. *άγκανος* < Koine Greek subst. *άκανος*, cf. MG dialectal *άγανος*, *άγκανος* 'awn, a stiff bristle'. See also ILNE, s.v. *άγανος* < Koine Greek subst. *άκανος* 'thistle-head' or 'a pine-thistle' LSJ). Cf. MG dialectal *άγανος*, *άγκανος*, meaning 'awn, a stiff bristle'. For the meaning of this word, see also Karanastasis (2003: 42).

55 Medieval *ἐξούριος* is interpreted by other scholars as *πονηρὸς* 'cunning, foxy' (see commentary by M. Crusius in Eideneier, *Spanos* 1977: 242).

eksurios is used also as an independent word elsewhere in Eideneier, *Spanos* 1977 (D 891 *«χαίροις μὲ τοὺς δαίμονας / τραγογένη, τριγένη, ἐξούριε»*).[56]

In general, the origin of such expressions (constructions and words thereof) lies in the heart of neologism, which, for the purposes of magic and the rhetoric inherent in it, are characterized by i) incomprehensible/indiscernible meaning and ii) characteristic sound patterns, of which repetition is the centre (see Passalis 2012). According to Passalis (2012: 12) "the majority [...] of the special vocabulary that we encounter in charms contains artificially constructed compounds". Passalis refers to cases of recasting a word in the form of a non-existing (but still semantically interpretable) compound, in juxtaposition to the original word, as in *merminga, proto-merminga* ('ant, chief-ant') and *kunupa, tri-kunupa* ('mosquito and thrice-mosquito') (Passalis 2012: 13). He adds that "the presence of these [constructions] in those genres is mainly connected to the mnemonic function of rhythm" (Abrahams 1968: 51; Sherzer 1990: 240, referred to in Passalis 2012: 13). Repetition and rhythm (as in music, mantra formulas etc.) has (psychosomatic) effects on the performer that become the prerequisites/conditions for the performance he needs to act. In this way, "sound is [...] transformed into an 'instrumental' tool for the modification of the extra-textual reality" (Sebeok 1974: 41 in Passalis 2012: 14), or, as Malinowski (1965: 218, 219 in Passalis 2012: 14) puts it, "sound in magic is a type of verbal *missile* replete with magic power".

To conclude, the exclusively Greek *X kse-X* pattern as a whole and the cross-linguistic echo-word construction *X [m]X* seem to exhibit strong commonalities. Not only do they have parallel semantic and pragmatic potential, but also seem to share common roots in possibly originating from the creative use of language in genres such as magical practice, spells, curses, language games, trickery and related situations, for strongly affective and rhetorical purposes.

2.3.3 Other reduplicative constructions involving prefixes and fixed morphemes

In Greek, a language with rich morphology, other prefixes may take part in reduplicative constructions with very similar functions to those explored in the previous sections. Reference is made to some of these cases below.

56 For more information about the semantics of these words see Eideneier, *Spanos* (1977: 294 and 313).

2.3.3.1 *X (ce) para-X*

The prefix *para-* comes from the Ancient Greek preposition *παρά* (*pará*), which meant "next to, near". In SMG, according to the Dictionary of Standard Modern Greek (DSMG), 1998 (Manolis Triantafyllidis Foundation), there are, in fact, two prefixes with the form *para-*; the first expresses – among other things – the notion of near similarity (*paraplisios* 'nearly similar', *paraiatrikos* 'paramedical'), existence/function that is parallel or outside the typical boundaries of what the base expresses (*parakratos* 'parastate', *parastratiotikos* 'paramilitary'), opposition (*paraloɣos* 'unreasonable', *paranomos* 'illegal') and deliberate distortion/ change of what the base expresses (*paraxaraso* 'counterfeit', *parerminevo* 'misinterpret'). The second *para-* expresses intensity and exaggeration, esp. with verbs, where it means 'overdo with something' (*paracimame* 'oversleep', *parafortono* 'overload').

Thus, similarly to polysemous *kse-*, that seems to give rise to two types of construction according to the basic meaning involved (removal/opposition or intensification), *para-* in SMG is involved into two different types of construction, as follows:

The first construction concerns mainly nouns and has the meaning "X and the like", with generally negative connotations about what is represented by *para-X* (or also by *X* in the case in question), as in examples (77)–(80):

(77) *eðo tosi ce tosi Solon, piites, parapiites, kalitexnes*
here so.many and so.many Solon poets *para*-poets artists
parakalitexnes, politici parapolitici, ipopolitici
para-artists politicians para-politicians hypo-politicians
ɣlifune apo ante na mi su po ce ti
lick.3PL from INTERJ PTCL not you.OBJ tell and what
ɣlifune ce exakoluθun ce iparxun [...][57]
lick.3PL and continue.3PL and exist.3PL
'Well, Solon, now so many poets and the like, artists and the like, politicians and the like, bad politicians lick...well, let me not tell you what they lick and continue to survive [...]'

(78) *me prixane oli i jiatri ci i parajiatri mi*
me swell.PFV.3PL all the doctors and the para-doctors not
vafis ta maʎa su ce mi vafis ta maʎa su
dye.IPFV the hair yours and not dye.IPFV the hair yours

57 Kostas Murselas, *Διά της ατόπου απαγωγής: Σατιρικοί διάλογοι* [ðia tis atopu apaɣojis: satirici ðialoji], Κέδρος (Kedros Publications), 1981, p. 64.

'They have tired me up, all the doctors and the like, telling me 'don't dye your hair' all the time'[58]

(79)

eci	*stin*	*Africi*	*tsakonondusan*	*panda*	*filés,*	*parafilés*	...[59]
there	in.the	Africa	fought.IPFV.3PL	always	tribes	*para*-tribes	

'There in Africa tribes of all sorts always used to fight with each other ...'

(80)

mesa	*sto*	*tetraðiaci*	*me*	*tis*	*simiosis*	[...]	*kuvalame*
in	the	notebook.DIM	with	the	notes		carry.3PL

ðjafora	*xartakça-*	*paraxartakça*[60]
various	paper.DIM.PL	*para*-papers.DIM.PL

'inside the notebook we carry various little papers of all sorts'

As is observed in the examples above, the construction may include *ce* 'and' (*X (ce) para-X*), but its presence is optional. Also syntactically speaking, X may be a whole Noun Phrase (as in 78, where the determiner is included). On the other hand, there are cases in which the two parts of the reduplicative pair are considered closely tied (closely enough to be connected with a hyphen, as in 80).[61]

With respect to semantics, note that, similarly to the equivalent *kse-* constructions discussed above, *para-* does not create existing lexemes within this reduplicative construction. That is, there are no lexemes such as *parapiitis* '*para*-poet', *parapolitikos* '*para*-politician.N'[62] or *parapaputsi* '*para*-shoe'. Once again, we are faced with occasional formations, which function as *ad hoc* synonyms of X (see 2.3.2 above), roughly with the meaning "something like X, but not as good as to be named X". That is, negative implications are also present here.

Note that negativity seems to arise through the meaning of *para-* per se: even outside reduplication, most of the meanings arising through *para-* prefixation have negative implications about the derivative word. More specifically, since the derivative is similar or parallel to the original, it is secondary, hence somehow inferior to the original. This is evident in words such as *parajos*

58 At http://tistrellis.pblogs.gr/2008/11/ta-eisagomena-mallia-ths-kefalhs-moy.html (last access January 5th 2017).

59 SKAI Radio, 20/04/2017.

60 At http://stardustia.blogspot.gr/2008_06_01_archive.html (last access January 5th, 2017).

61 As was discussed in Kallergi (2015), the use of the hyphen in reduplication (as far as written speech is concerned) seems to vary according to speakers' subjective judgment or preference.

62 Note that *parapolitikos* in MG is an adjective meaning "the one relating to the backstage in politics". However, within the reduplicative construction in the relevant example, *parapolitici* is a noun, therefore the formal identity with the adjective must be interpreted as accidental.

'*para*-son' 'helper, apprentice', *paramajiras* '*para*-cook' 'vice chef', *paraðuleftra* '*para*-worker' "servant" etc.

The second construction that involves *para-* can be represented, at least in SMG, as *V ce paraV*, since it applies only to verbs and expresses intensification and/or strong assertion (example 82). In the Pontic dialect, one encounters *X ce para-X*, where *X* is either a verb or an adverb (example 81). Notably, all the *para-X* constituents in our source of Pontic data (Papadopoulos 1958) are lexical entries themselves.

(81)	*anoma*	*ce*	*paranoma*	*pis*	*tin ðulían*
	non-legal.ADV	and	illegal.ADV	do.2SG	the job

'You do the job most illegally' [Papadopoulos 1958, s.v. *παράνομα*]

(82)	*an*	*ce*	*i ðicaiosini*	*ine*	*tifli,*	*orismenes*	*fores*	*vlepi*	*ce*
	if	and	the justice	is	blind	some	times	see.3SG	and
	paravlepi ...[63]								
	para-see.3SG								

'Although justice is blind, sometimes it sees and very well indeed ...'

The choice of the construction in (82) seems to be based on rhetorical grounds: the word form *paravlepo* is an existing lexeme in MG, meaning 'deliberately overlook'. In this case, it coincides with the emphatic assertion of the statement that justice is not blind at all: not only does it see, but it also overlooks injustice. However, very often in SMG the *para-* derivative in the *V ce paraV* construction is an occasional word. For instance, *paraɣustaro* as in (83) does not exist independently in the MG lexicon.

(83)	*les*	*xazomares!*	*ɣustari*	*ce*	*paraɣustari!*	*apla*
	say.2SG	nonsense	fancy.3SG	and	para-fancy.3SG	simply

	ðen	*bori*	*na*	*to*	*ekfras-i*	*elefθera* [...][64]
	not	can.3SG	PTCL	it.OBJ	express.PFV-3SG	freely

'You talk nonsense! Of course he wants! He just can't express it openly [...]'

Also, the *para-* derivatives in this construction may be homophonous to the derivatives consisting of a verb plus the second prefix *para-* mentioned above, that with the meaning of exaggeration. For example, we saw *paraime* being

63 At http://www.avgi.gr/article/10811/7756978/otan-e-dikaiosyne-parablepei (last access January 6th, 2017).

64 At www.forums.gr (comment posted on October 29th, 2012; last access January 6th, 2017).

interpreted as "I am indeed" in (76) above (2.3.2). However, *paraime* outside reduplication means "I am too much", as in:

(84)	*otan*	*mia*	*prosfora*	*paraine*	*kali*	*ja*	*na*	*ine*	*aliθini*[65]
	when	an	offer	*para*-be.3.SG	good	to	PTCL	be.3SG	real

'when an offer is too good to be true'

As the English translations imply, *paraime* in the reduplicative construction has positive implications (at least, strong assertion in opposition to a previous statement), whereas the existing, independent *para*-derivatives express excess, which is rather negatively appreciated. Yet, the verbs express exaggeration in both cases.

Thus, once again, the situation is very similar to *kse*-reduplication: two distinct constructions, one having the meaning "X and the like" plus negative connotations and the other being emphatic/intensifying/assertive, mostly applying to verbs and having possibly also positive implications.

2.3.3.2 *X (ce) andi-X*

In the centre of the semantics of the prefix *andi-* 'anti-' is opposition and antithesis (as, e.g., in *andipeðaɣojikos* 'anti-pedagogical' and *andixristos* 'antichrist'). However, similarly to the previously mentioned prefixes, *andi-* is highly productive and polysemous, expressing – among other things – lack of certain characteristics (*andiiroas* 'anti-hero'), reciprocal movement (*andekðicisi* 'retaliation'), equivalence (*andistixos* 'equivalent') – comparison (*andiparuvalo* 'compare/ contrast'), counterbalance (*andivaro* 'counterweight'), substitution (*andiproeðros* 'vice-president'), and very marginally, intensification, as in *andíθama* ('anti-miracle' "great miracle"). In fact, *andíθama* sounds literary and is probably only encountered in the reduplicative construction *θáma c' andíθama* 'miracle and anti-miracle' "the greatest of miracles".

Somewhat more productively in SMG (but not as much as *kse-*), the prefix is involved in reduplicative constructions of the 'X and the like' type:

(85)	*avrio*	*ta*	*lete*	*ola.*	*Kritices*	*andikritices, klp.*[66]
	tomorrow	them	say.2PL	everything	critiques	*andi*-critiques etc.

'You'll talk tomorrow. About critiques and all.'

65 At http://www.insuranceforum.gr/forum/viewtopic.php?f=106&t=613&start=0 (last access January 6th, 2017).

66 The instance was attested in Athens (taken from Magdalene Konstantinidou's personal record).

(86)

[...],	*analisis,*	*kritices,*	*andikritices,*	*pu*	*kanate*
	analyses	critiques	*andi*-critiques	which	do.2PL.PFV.PST

s'	*o,ti*	*afora*	*to ɵezmo* [...][67]
on	whatever	concern.3SG.PR	the institution

'...analyses, critiques and such, which you have made with respect to the institution [...]'

The reappearance of this construction with the lexeme *kritici* 'critique' might suggest the existence of a fixed expression. Nevertheless, there is no such expression in dictionaries, nor is *andicritici* 'andi-critique' a lexical entry. In fact, it may be the case that *andi-* carries a [+literary] or [+educated] feature that blocks it from appearing with everyday common nouns, such as, e.g., *karekles* 'chair.PL'.

As mentioned above, speakers of SMG are perhaps only familiar with the intensifying/emphatic use of *andi-* in *ɵáma c' andíɵama* 'miracle and anti-miracle' "the greatest of miracles". In contrast, Modern Greek dialects exhibit a large number of expressions of the type *X ce andi-X*, with an emphatic function (see examples 87–90 below, all drawn from ILNE). Note that also this construction carries a literary flavor (most of the instances below come from folk songs, riddles and fixed forms of language). But the process seems productive in applying to a variety of lexical classes (nouns, verbs and adverbs). Similarly to emphatic constructions with other prefixes, *X ce andi-X* seems to necessarily involve *ce/ci* 'and' (an exception is example 88).

(87)

kitaza	*krifa*	*ci*	*andikrifa*	*na*	*si*	*peripiumi*
look.1SG.IPFV	secretly	and	*andi*-secretly	PTCL	you	take.care.1SG

ce	*si*	*tora*	*mi*	*ðçoxnis* (Thessaly)[68]
and	you	now	me	turn.way.2SG.PR

'I was trying to take care of you with most discretion, and now you are turning me away'

(88)

mu	*mina*	*m'*	*andimina*	*na*	*kamume*
me	ask.3SG.PR	me	*andi*-ask.3SG.PR	PTCL	do.1PL.PR

sibeɵerjá (Athens)[69]
marriage.match

'It keeps asking me all the time to involve in marriage matches'

67 At http://www.efivoi.gr/site_stuff/files/1997/praktika_eidikh_synedriash_olomeleia.pdf (last access January 9th, 2017).

68 This is most probably a fixed expression, appearing in this particular example, as well as in a folk song coming from the same region (Thessaly).

69 Taken from a folk song.

(89) *imun* *pedcios* *ci* *andipedcios* [...] (Cyprus: Pafos)[70]
be.1SG.IPFV youngster and *andi*-youngster
'I was a strong young man [...]'

(90) *efaje* *ksiilo* *ce* *adiksilo* (Peloponnese: Lakonia)[71]
eat.3SG.PFV.PST wood and *andi*-wood
'S/he was beaten hard'

Thus, MG in total (SMG and dialects) exhibits relatively low productivity with respect to reduplicative constructions with *andi-*. Notably, however, the standard variety seems to differentiate itself from dialects in using a productive 'X and the like' pattern with *andi-*, instead of an emphatic (or, even, pleonastic) pattern (which may also have lexical representatives).

Note, in closing, that a similar situation seems to be the case with the prefix *apo-*, which has a similar range of meanings with the prefixes discussed so far. There is abundance of evidence on the use of emphatic/pleonastic expressions of the *X ce apo-X* type in dialects. Nevertheless, SMG exhibits only one expression of this type and this is a frozen expression (a lexical entry): *skata ci apóskata* 'shit and *apo*-shit' "(in) deep shit" (for the emphatic assertion of a bad condition or a failure). Given that a productive reduplicative pattern with *apo-* is missing from the standard variety, we will omit data from the dialects as well.

2.3.3.3 *X (ce) skat-x*

The final case to be discussed is *skat-*, which is marginal in two senses. First, referring to a taboo element ('shit'), it belongs to marginal/slang vocabulary. Second, it is not a prefix; the MG prefixoid *skato-* 'shitty-' (which expresses rejection, indignation or annoyance, see e.g., Christopoulou 2016) has incorporated the linking morpheme *-o-* (i.e., the element that links roots in MG compounding, often called the *compound index*, Ralli 2005: 165) and is along the way towards grammaticalization from base to affix (Giannoulopoulou 2006: 278–279, Vounchev 2016, Christopoulou 2016). The morpheme *skat-*, on the other hand, does not include the linking morpheme *-o-* in the reduplicative constructions discussed below. In this sense, *skat-* is most probably the lexical root of the lexeme *skat-ó* ('shit-N.SG').

70 Taken from a folk song.
71 The expression *troo ksilo* 'to eat wood' means "to be beaten hard" also in SMG. The reduplicative expression, however, appears only in dialectal speech.

However, the morpheme *skat-* is rather special, because its meaning in the environments encountered below is much closer to the semantic/functional potential of the prefixoid *skato-*. That is, the meaning of the root *skat-* has been extended to cases where no reference is made to human excrement, but strong pragmatic implications arise (Christopoulou 2016). Form-wise, its combinatorial potential has been restricted, since, as we see below, its use is confined to combinations with derivational suffixes only. In this sense, it is not purely a root, but can be considered a fixed morpheme, serving the creation of neologisms of a particular kind and structure.

In lack of evidence from dialects, we will refer to the use of this morpheme in highly informal contexts of discourse in SMG and its involvement in reduplicative constructions of the following type:

(91) [...] *mazevun* *o,ti* *petraðaci* *ce* *skataci* *boris*
pick.PL.PR whatever stone.DIM and shit.DIM can.2SG.PR
na *fadastis*[72]
PTCL imagine.2SG
'[...] whatever kind of little stone and stuff you can imagine sticks to them [bicycle tires of a specific brand]'

(92) *xrisimopiis* *ena* *epimetalomeno* *plastiko* *opos* *afto*
use.2SG.PR a metalized plastic such.as this
sta *patataça-pitsiɲa-skatiɲa* *ce* *kaθarises*[73]
in.the chips pitsinia[74] skatinia and clean.2SG.PFV.PST
'You use a metalized plastic such as those found in chips, *pitsinia* and such stuff and you are done'

(93) *"ax* *savataci* *θa* *vɣo* *me* *to* *pareaci*
INTERJ Saturday.DIM FUT go.out.1SG with the company.DIM
sto *clabaci"*
to.the club.DIM
(ai *mori* *vlameni* *"savataci* *pareaci* *ce* *skataci")*[75]
INTERJ INTERJ idiot Saturday.DIM company.DIM and shit.DIM
'"Oh, it's (little) Saturday and I'm going to go out at the (little) club with my (little) company" (give me a break, you "little Saturday, little company and shit" idiot)'

72 At http://www.podilates.gr/forum/agora-podilatoy-klp/cube-aim-2011-gnomes-gia-sygkekrimeno-podilato (last access January 12th, 2017).

73 Accessed on Google as "πιτσίνια σκατίνια" (last access January 11th, 2017).

74 The word is a Greek brand name for a type of crisps.

75 At http://www.e-steki.gr/showthread.php?t=5428&page=11 (last access January 12th, 2017).

(94) *ala* *o* *ʎiondarakos* *θa* *jini* *pali* *leptulis* *ce*
but the lion.DIM.FUT FUT become.3SG again slim.DIM and
skatulis [...][76]
shit.DIM
'but Liondarakos[77] will become again slim and bullshit [...]'

(95) *Sað-istis* *ce* *skat-istis*[78]
sad-ist and shit-ist
'Sadist schmadist'

At a first glance, the expression *ce skataci* 'and shit.DIM' as such may pass as a filler phrase or a discourse marker (of rejection, indignation etc.) or an expletive attributive (see Huddlestone & Pullum 2002). However, the fact that *ce skataci* appears in juxtaposition to *petraðaci* and *pareaci* in strongly quotative contexts cannot be ignored as accidental. Also, in the rest of examples, the copy-word rhymes with the base-word, carrying *skat-* at its onset and keeping the last syllables intact (*skat-inia*, *skat-ulis*, *skat-istis*). Under this light, it seems more plausible to say that we are faced with an echo-construction. In fact, the construction in question could be a case of what is known as *Melodic Overwriting*.

Melodic Overwriting is a term used by McCarthy & Prince (1986), Yip (1992), Alderete et al. (1999) and others to refer to cases of reduplication, whereby the copy exhibits phonological material (typically, in its onset) that seems to be replacing material of the original word (see also Inkelas & Zoll 2005, Goldsmith et al. 2011). Thus, echo-constructions such as the Turkish *mühleme* (*kitap mitap* 'books and the like') and the American English *schm-* reduplication (*fat schmat* 'what if she's fat') are often analyzed as cases of Melodic Overwriting.[79]

76 Accessed on Google as "*λεπτούλης και σκατούλης*" (last access January 12th, 2017).

77 This is probably a nickname of a person engaged in the blog discussion, who has stated that he is going to lose weight and regain his built body after finishing his duty in the army. The sentence is an answer by another blogger, who expresses irony and disbelief towards that statement.

78 At http://www.redhoops.gr/forum/showthread.php?p=222714 (last access January 16th, 2017).

79 In fact, in sources such as Inkelas & Zoll (2005) and Goldsmith et al. (2011), reference is made not only to processes of replacement, but also to processes of addition of material, as is exactly the case when there is no consonantal onset in the original word in echo-word constructions (MG *ozes mozes* "nail enamels and such", Am. Eng. *OT-schmOT*). In this sense, reduplicative constructions involving prefixes, such as the ones discussed here, could perhaps also be considered cases of melodic overwriting (i.e., fixed segmentism of a "morphological type", Alderete et al. 1999). However, unlike the usual case with fixed segmentism in other languages, the MG prefixal morphemes in question appear elsewhere in the language (viz. they

In the case of *skat-*, the part to be overwritten seems to be the root (the lexical/content part) of the original word. This is replaced by the root *skat-*, leaving only the "ending" (i.e., the derivational and inflectional suffixes) of the target word. In this way, the copy "sounds like" (echoes) the target word. In fact, the copying material (sometimes including *ce* 'and') rhymes with the original word, in carrying the same number of syllables. For example, in example 93, [pa-re-a-ci] and [ce-ska-ta-ci] both have four syllables. However, in other cases, base and copy share the same number of syllables without taking *ce* into account. In this case, *ce* seems to interrupt the rhyming sequence of base and copy (examples 94–95). It appears that, similarly to constructions with prefixes above, *ce* may be an integral part of the construction. For instance, example (96) would sound incomplete without *ce*:

(96)	*?ala*	*o*	*ʎiondarakos*	*θa*	*jinni*	*pali*	*leptulis*	*skatulis* [...]
	but	the	lion.DIM	FUT	become.3SG	again	slim.DIM	shit.DIM

'?but Liondarakos will become again slim bullshit [...]'

One may assume that the presence of *ce* in reduplication relates to the quotative function of word repetition, even if the word that is copied does not reappear in discourse as such (see 2.3 for an analogous *X [m]X* construction, and Kakridi-Ferrari 1998, Kallergi 2015 for similar observations on pragmatic repetition). Also, the overall sequence of words in X *skat-X*, as in typical echo-word constructions in MG, often reflects the notion of a group ("X and Y (=the like)) requiring the presence of *ce*. Still, the phonological role of *ce* and the issue of whether it can be accepted as part of constructions characterized by Melodic Overwriting needs further examination. For the time being, it is important to note that the presence of *ce* marks the status of this reduplicative construction as non-prototypical. Another factor that plays a role is of course the formal non-identity between base and copy. Also as a case of melodic overwriting, it is non-typical because, unlike the most famous cases discussed under this rubric, *skat-* carries meaning in itself and tends to create neologisms that, although they are highly unlikely to appear outside reduplication, can be somehow interpreted on the basis of the combination between *skat-* and the derivational suffixes involved.[80]

are proper prefixes). Also, unlike typical Melodic Overwriting, in reduplicative constructions with *kse-*, *para-* etc. the prefix never replaces phonological material of the base.

80 This goes especially for *skat-* derivatives such as *skatini* 'shit.DIM' and *skatistis* 'shit.ist', that do not coincide with existing words (such as *skataci* 'shit.DIM' or *skatulis* 'shit.DIM').

3 Conclusion

In this paper we have examined different patterns of reduplication involving morphophonological variation/distortion, namely, the *mühleme* or echo-word construction and reduplication with prefixation by *kse-*, *para-*, *anti-* and *skat-*. Summing up the analysis, one may observe the strong and highly interesting parallels between different patterns of reduplication with variation/distortion: almost all reduplicative (or TRCV) constructions that have been discussed exhibit a two-fold development, or, as a matter of fact, a semantic distinction, which is also formally designated: on the one hand, they can all express the meaning 'X and the like'; on the other hand, they may be all used for the expression of emphasis/exaggeration/strong assertion and/or greater intensity. In the latter case, the reduplicative patterns in question often involve the appearance of the conjuction *ce* 'and' between the constituents of the expression. Note that *ce* may also appear in constructions of the 'X and the like' type. However, in emphatic constructions its presence seems to be obligatory.

For this and other reasons relating to the semantic and formal characteristics of the emphatic constructions of the *X ce yX* type (which differentiate them from their 'X and the like' peers), one may assume that the former move further away from an assumed prototype of reduplication and towards the realms of (lexical) juxtaposition and/or pragmatic repetition. However, in the absence of supporting evidence for the productivity (or unproductivity) of certain patterns, the above observation is far from a definite conclusion.

Regarding the pragmatics of the types of non-prototypical reduplication undertaken here, a central notion that seems to emerge is negativity. More interestingly, even, negative connotations (among them pejorativity, contempt and rejection) seem to run in parallel with (otherwise positive) pragmatic results, such as ludicness and playfulness. Other concepts, such as the grouping of inhomogeneous (yet similar) entities and the concomitant negative implications on the quality of the copy (or the group of an original entity and its copy(ies)) seem to arise. In all these cases, it seems plausible to assume an iconic connection between linguistic, formal non-identity/non-homogeneity and situations where non-linguistic, real-life entities are involved. Thus, earlier assumptions on this connection (e.g., Konstantinidou 2004, Stolz 2008) are supported by the evidence brought in this paper.

Finally, with respect to the origin of reduplicative constructions with variation/distortion, it has been noted at various points that rhetoric of a specific kind has probably played an important role: at least regarding the major reduplication types examined (i.e., [m]-reduplication and *kse*-reduplication), the genesis of such constructions seems to be found in practices of verbal magic and the performance of verbal acts of harming, deceiving or concealing. This assumption helps towards a unified analysis between the currently productive patterns of distortive reduplication and a broad category of frozen expressions consisting of near-identical non-words. Apart from the formal similarities between the two types (productive and unproductive), it may be assumed that the latter – originating in magic, where words typically have no (specifiable) meaning – may have given rise to productive patterns of reduplication. On the other hand, the fact that phonological distortion is found in reduplicative ideophones with similar semantic/pragmatic repercussions perhaps signifies a highly natural and possibly universal tendency towards patterns of morphophonological distortion and/or near-identity, i.e., patterns of non-prototypical reduplication.

Abbreviations

1, 2, 3	first, second, third person
ACC	accusative
ADV	adverb
AUG	augmentative
AUX	auxiliary
CLIT	clitic
DAT	dative
DIM	diminutive
DSMG	Dictionary of Standard Modern Greek
ECHO	echo
FEM	feminine
FUT	future
GEN	genitive
IMP	imperative
INF	infinitive
INTERJ	interjection
IPFV	imperfective
MASC	masculine
MG	Modern Greek
MS	manuscript
N	neuter
NEG	negative
NOM	nominative
OBJ	object
PASS	passive
PFV	perfective
PL	plural

PR	present
PRF	perfect
PRS	present
PTCL	particle
PST	past
PTCPL	participle
SG	singular
SMG	Standard Modern Greek
S.V.	sub verbo
TRVC	total reduplication cum variation
UMGD	Utilitarian Modern Greek Dictionary
VOC	vocative

References

Abrahams, Roger D. 1968. A rhetoric of everyday life: Traditional conversational genres. *Southern Folklore Quarterly* 32. 44–59.

Alderete, John, Jill Beckman, Laura Benua, Amalia Gnanadesikan, John J. McCarthy & Suzanne Urbanczyk. 1999. Reduplication with fixed segmentism. *Linguistic Inquiry* 30. 327–364.

Anastasiadis, Vassilios K. 1976. *Η σύνταξη στο φαρασιώτικο ιδίωμα της Καππαδοκίας* [Syntax in the local dialect of Farasa Cappadocia]. Θεσσαλονίκη.

Andriotis, Nikolaos. 1951. *Ετυμολογικό λεξικό της κοινής Νεοελληνικής* [Etymological dictionary of Standard Modern Greek]. Αθήναι: Institut Français d'Athènes, 3rd ed. Θεσσαλονίκη: ΙΝΣ, 1983.

Archive ILNE = Αρχείο του Κέντρου Ερεύνης των Νεοελληνικών Διαλέκτων και Ιδιωμάτων – ΙΛΝΕ της Ακαδημίας Αθηνών [Archive of the Research Centre for Modern Greek Dialects – ILNE of the Academy of Athens].

Athanasiadis, Stathis. 1977. *Συντακτικό της ποντιακής διαλέκτου* [Syntax of Pontic dialect], Καστανιά Βεροίας. https://pontiakilelapa.wordpress.com/.../ποντιακός-λόγος-σύνταξη-γραμματική.

Babiniotis, George. 1998. *Λεξικό της νέας ελληνικής γλώσσας, με σχόλια για τη σωστή χρήση των λέξεων* [Dictionary of Modern Greek language...]. Αθήνα: Κέντρο Λεξικολογίας, 2nd ed. 2005.

Bauer, Laurie. 1988. *Introduction to linguistic morphology*. Edinburgh: Edinburgh University Press.

Borsje, Jacqueline. 2011. Digitizing Irish and Dutch charms. In Tatyana A. Mikhailova, Jonathan Roper, Andrey L. Toporkov & Dmitry S. Nikolayev (eds.), *Oral charms in structural and comparative light*, 128–137. Moscow: PROBEL.

Chioti, Alexandra. 2010. Οι παγιωμένες εκφράσεις της Νέας Ελληνικής: Ιστορική διάσταση, ταξινόμηση και στεροτυπικότητα. Διδακτορική διατριβή. [Frozen expressions in Modern Greek: Historical dimension, classification and stereotypicality. PhD thesis]. Αριστοτέλειο Πανεπιστήμιο Θεσσαλονίκης/Aristotle University of Thessaloniki.

Christopoulou, Katerina. 2016. Μια λεξικολογική προσέγγιση στο περιθωριακό λεξιλόγιο της Νέας Ελληνικής [A lexicological approach to the marginal vocabulary of Modern Greek]. Unpublished PhD thesis. University of Patras.
Dawkins, Richard McGillivray. 1916. *Modern Greek in Asia Minor. A study of the dialects of Silli, Cappadocia and Pharasa with grammar, texts, translations and glossary*. Cambridge: University Press.
Dingemanse, Mark. 2012. Advances in the cross-linguistic study of ideophones. *Language and Linguistics Compass* 6(10). 654–672.
DSMG 1998 = *Λεξικό της Κοινής Νεοελληνικής* [Dictionary of Standard Modern Greek]. Θεσσαλονίκη: ΙΝΣ [Institute for Modern Greek Studies].
Efthymiou, Angeliki. 2001. Το Νεοελληνικό Πρόθημα ξε-: οι Έννοιες της Απομάκρυνσης και της Αλλαγής Κατάστασης [The Modern Greek prefix *kse-*: the concepts of removal and change of state]. *Studies in Greek Linguistics* 2000: 202–213.
Efthymiou, Angeliki. 2002. Σημασιολογικές Παρατηρήσεις για τα Νεοελληνικά Προθήματα ξε-, εκ- από- [Semantic observations on the Modern Greek prefixes *kse-*, *ek-*, *apo-*]. *Studies in Greek Linguistics* 2001: 199–209.
Efthymiou, Angeliki. 2003. Προθήματα ή α' συνθετικά που δηλώνουν επίταση στα Νέα Ελληνικά [Prefixes or first compound constituents that denote intensity in Modern Greek]. *Proceedings of the 23rd Annual Meeting of the Department of Linguistics, School of Philology, Aristotle University of Thessaloniki*, 519–528.
Eideneier, *Spanos*. 1977. = Eideneier, Hans. *Spanos: eine byzantinische Satire in der Form einer Parodie. Einleitung, kritischer Text, Kommentar und Glossar besorgt von Hans Eideneier* [Supplementa Byzantina 5]. Berlin: De Gruyter.
Eideneier, *Spanos*. 1990. = Eideneier, Hans (ed.). *Σπανός*.[Spanos] Αθήνα: Ερμής.
Evangelatos, Spyros (ed.). 2002. *Δ. Κ. Βυζάντιος. Η Βαβυλωνία*, γ' έκδ. [D. K. Byzantios. Babylonia [1836] 3rd edition]. Αθήνα: Εστία.
Fleischer, Wolfgang. 1982. *Phraseologie der deutschen Gegenwartssprache*. Leipzig: VEB Bibliographisches Institut.
Fraser, Bruse. 1970. Idioms within a transformational grammar. *Foundations of Language* 6. 22–42.
Giannoulopoulou, Giannoula. 2006. Dialectological research and linguistic theory: The case of compounding. In Mark Janse, Brian Joseph & Angela Ralli (eds), *Proceedings of the 2nd International Conference of Modern Greek Dialects and Linguistic Theory*, 68–82. Patras: University of Patras.
Goldsmith, John A., Jason Riggley & Alan L.C. Yu (eds.). 2011. *The handbook of phonological theory*. West Sussex/Oxford: Wiley-Blackwell.
Hatzidakis, Georgios N. 1905. *Μεσαιωνικά και Νέα Ελληνικά*. τ. Α', 1905, τ. Β', 1907 [Medieval and Modern Greek, vol. 1, 1905, vol. 2 1907]. Αθήναι.
Hatzidakis, Georgios N. 1914. Περί της χρήσεως της προθέσεως *εκ*, *εξ* (*ξε-*) εν τη νεωτέρα Ελληνική. [On the use of the preposition *ek*, *eks* (*kse-*) in Modern Greek]. *Λεξικογραφικόν Αρχείον της Μέσης και Νέας Ελληνικής* 1. 8-47.
Hatzitaki-Kapsomenou, Chrysoula. 2001. *Θησαυρός νεοελληνικών αινιγμάτων*, β' έκδ. [Thesaurus of Modern Greek Reddles, 2nd edition]. Ηράκλειο: Πανεπιστημιακές Εκδόσεις Κρήτης (α' έκδ. 2000). http://tobias-lib.uni-tuebingen.de/volltexte/2003/809/pdf/AA_turedu_diss_130603.pdf.
Hinton, Leanne, Johanna Vichols & John J. Ohala. (eds.). 1994. *Sound symbolism*. Cambridge: Cambridge University Press.

Huddlestone, Rodney & Geoffrey Pullum. 2002. *The Cambridge grammar of the English language*. Cambridge: Cambridge University Press.
ILNE 1933– = *Ιστορικόν Λεξικόν της Νέας Ελληνικής, της τε κοινώς ομιλουμένης και των ιδιωμάτων*, 1-6 (α-διάλεκτος) [*Historical Dictionary of Modern Greek, both of the common spoken language and the dialects*]. Athens: Academy of Athens.
Inkelas, Sharon & Cheryl Zoll. 2005. *Reduplication: Doubling in morphology*. Cambridge: Cambridge University Press.
Johanson, Lars. 2002. *Structural factors in Turkic language contacts*. Richmond MD: Curzon.
Joseph, Brian D. 1985. European Hellenism and Greek nationalism: Some effects on Greek linguistic scholarship. *Journal of Modern Greek Studies* 3. 87–96. www.ling.ohio-state.edu/.../1985hellenism.pdf.
Joseph, Brian D. & Irene Philippaki-Wartburton. 1987. *Modern Greek*. London/Wolfeboro, N. H.: Croom Helm.
Kakridi-Ferrari, Maria. 1998. Επανάληψη: η λειτουργία της ως βασικού μηχανισμού στη γλώσσα [Repetition: Its function as a basic mechanism in language]. Phd Dissertation, University of Athens.
Kallergi, Haritini. 2013. The role of repetition in the rise of concessivity. In Nikolaos Lavidas, Thomai Alexiou & Areti Sougari (eds.), *Major trends in theoretical and applied linguistics: Selected papers from the 20th International Symposium on Theoretical and Applied Linguistics, April 1–3, 2011*, 319–334. London: Versita.
Kallergi, Haritini. 2015. *Reduplication at the word level: The Greek facts in typological perspective*. Berlin/New York: Mouton de Gruyter.
Karanastasis, Tasos. A. 2003. Ακολουθία του Ανοσίου Τραγογένη Σπανού. Χαρακτήρας και χρονολόγηση. Μία ερμηνευτική προσέγγιση. [The divine service of the unholy goat bearded milky-bearded]. PhD thesis, Aristotle University of Thessaloniki.
Keane, Eleanor. 2005. Phrasal reduplication and dual description. In Bernhard Hurch (ed.), *Studies on reduplication*, 239–262. Berlin: Mouton de Gruyter.
Kikvidze, Zaal. 2016. How to gloss and translate Georgian echo-paired words. In Artur G. Hudmanyana & Serfiy I. Sidorenko (eds.), *General and specialist translation/ interpretation: Theory, methods, practice. International conference papers*, 168–172. Kyiv, Ukraine: AgrarMedia Group.
Knobloch, Johann. 1989. Ein lateinischer Zauberspruch bei Marcellus Empiricus. *Rheinisches Museum für Philologie* 132. 408–409.
Konstantinidou, Magdalene. 1997. *Sprache und Gefühl. Semiotische und andere Aspekte einer Relation*. Hamburg: Buske.
Konstantinidou, Magdalene. 2004. Παραμορφωτική αναδίπλωση λέξεων. Ένα διαλεκτικό σχήμα λόγου και ο λειτουργικός του ρόλος [Distortive lexical reduplication. A dialectical rhetorical figure and its functional role]. *Μελέτες για την Ελληνική Γλώσσα 24. Πρακτικά της Ετήσιας Συνάντησης του Τομέα Γλωσσολογίας του Τμήματος Φιλολογίας της Φιλοσοφικής Σχολής του Α.Π.Θ. (Θεσσαλονίκη, 9–11 Μαΐου 2003)*, 345–356. Θεσσαλονίκη.
Konstantinidou, Magdalene. 2005. Degenerative lexikalische Reduplikation durch [m] im Anlaut: Eine eigenartige rhetorische Figur im Mundartgriechischen und Türkischen. In Eleni Butulussi, Evangelia Karagiannidou & Katerina Zachu (eds.), *Sprache und Multikulturalität. Festschrift für Professor Käthi Dorfmüller-Karpusa*, 253–274. Thessaloniki: University Studio Press.
Konstantinidou, Magdalene & Ekaterini Tzamali. 2012. Οι πηγές του Ιστορικού Λεξικού της Νέας Ελληνικής της Ακαδημίας Αθηνών [The sources of the Historical Dictionary of Modern

Greek of the Academy of Athens]. *Λεξικογραφικόν Δελτίον της Ακαδημίας Αθηνών* 26. 279–381.

Kopidakis, Michalis. 2007. Η σάρρα, η μάρρα και το κακό συναπάντημα [The *sarra*, the *marra* and the bad encounter]. In Εβίνα Σιστάκου (επιμ.), *Μιχάλης Ζ. Κοπιδάκης. Μελετήματα*, 250–252. Αθήνα: Πατάκης.

Kriaras, Emmanuel. 1968–. *Λεξικό της μεσαιωνικής ελληνικής δημώδους γραμματείας, 1100–1669*, 1–20 (*α-σταματώ*) [*Dictionary of vernacular medieval Greek literature, 1100–1669*]. Θεσσαλονίκη: Κέντρο Ελληνικής Γλώσσας.

Kriaras, *Epitome* = Ι. Ν. Καζάζης, Τ. Α. Καραναστάσης, Ε. Ζέρβου, Μ. Κεσόγλου, Χ. Πασσαλής, Π. Μαυρομματίδου & Α. Δεληγιαννοπούλου. *Επιτομή του λεξικού της μεσαιωνικής ελληνικής δημώδους γραμματείας 1100–1669 του Εμμανουήλ Κριαρά* [Epitome of the Dictionary of vernacular medieval Greek literature 1100–1669 of Emmanuel Kriaras]. Θεσσαλονίκη: Κέντρο Ελληνικής Γλώσσας, τ. Α' 2001, τ. Β' 2003.

Krulak, Todd. 2016. Introduction: Defining competition in Neoplatonism. In Nathaniel P. DesRosiers & Lily C. Vuong (eds.), *Religious competition in the Greco-Roman world*, 79–84. Atlanta: SBL Press.

Laographia (= *Λαογραφία*). Bulletin de la Société Hellénique de Laographie (Folklore – Ethnographie – Vie sociale) Athènes 1909–.

Lelekos, Michael. 1888. *Επιδόρπιον Β'* [*Dessert*]. Αθήναι.

Levy, Harry. 1980. An Anatolian language-trait in Byzantios' *Babylonia* and parallel traits on three continents. *Modern Greek Studies Association (MGSA) Bulletin* 12. 47–55.

Loukopoulos, Dimitrios. 1938. *Γεωργικά της Ρούμελης* [Agricaltural issues of Central Greece]. Αθήναι.

Malinowski, Bronislaw. 1922. *Argonauts of the western pacific: An account of native enterprise and adventure in the archipelagous of Melanesian New Guinea*. New York: E. P. Dutton.

Marcelli de medicamentis liber. 1968. Marcellus über Heilmitter, In M. Niedermann & E. Liechtenhan (eds.) *Corpus Medicorum Latinorum* 5, 2d ed. Berlin: Akademie Verlag.

Mauss, Marcel & Henri Hubert. 1902. Esquisse d'une théorie général de la magie. *Année Sociologique* 7. 1–146 (édition électronique 2002: http://classiques.uqac.ca/classiques/mauss_marcel/socio_et_anthropo/1_esquisse_magie/esquisse_ magie.pdf).

McCarthy, John J. & Alan Prince. 1986. Prosodic morphology. Unpublished ms. University of Massachusetts, Amherst and Brandeis University.

Mel'čuk, Igor. 1996. *Cours de morphologie générale. Vol. 3. Troisième partie: Moyens morfologiques. Quatrième partie. Syntactiques morfologiques*. Montréal: Presses de l'Université de Montréal.

Minkova, Donka. 2002. Ablaut reduplication in English: The criss-crossing of prosody and verbal act. *English Language and Literature* 6. 133–169.

Müller, Hans-Georg. 2004. *Reduplikation im Türkischen: Morphophonologische Untersuchungen*. Wiesbaden: Harrassowitz. http://tobias-lib.uni-tuebingen.de/volltex-te/2003/809/ pdf/AA_turedu_diss_130603.pdf

Papadopoulos, Anthimos. 1953. Φρασεολογικά [Phraseology]. *Λεξικογραφικόν Δελτίον* 6, 3–92.

Papadopoulos, Anthimos. 1958. *Ιστορικόν Λεξικόν της Ποντικής Διαλέκτου* [Historical dictionary of the Pontic dialect]. Αθήναι: Επιτροπή Ποντιακών Μελετών.

Papanastasiou, Georgios. 2012. The preverb ξε- in Standard Modern Greek and in Modern Greek dialects. In Zoe Gavriilidou, Angeliki Efthymiou, Evangelia Thomadaki & Penelope Kambakis-Vougiouklis (eds), *Selected papers of the 10th International Congress of Greek Linguistics*, 492-498. Komotini/Greece: Democritus University of Thrace.

Passalis, Haralampos. 2012. From the power of words to the power of rhetoric: Nonsense, pseudo-nonsense words, archaisms and artificially constructed compounds in Greek oral charms. *Incantatio (An International Journal on Charms, Charmers and Charming)* 2. 7–22.

Politis, Nikolaos. 1900. *Παροιμίαι, τ. 2.* [Sayings]. Αθήναι: Βιβλιοθήκη Μαρασλή.

Potts, Christopher. 2007. The expressive dimension. *Theoretical Linguistics* 33(2). 165–198.

Psaltes, Stamatios B. 1915. Ετυμολογικά [Etymological issues]. *Αθηνά* 27. *Λεξικογραφικόν Αρχείον* 2, 98–114.

Ralli, Angela. 2005. *Μορφολογία* [Morphology]. Athens: Patakis.

Rangavis, Alexandros Rizos. 1874. *Άπαντα τα Φιλολογικά, τ. 3* [Collected literary writings]. Αθήναι.

Sarantakos, Nikos. 2013. Κάνε μου λιγάκι μμμ [Do me some *mmm*]. In Nikos Sarantakos' website *Οι λέξεις έχουν τη δική τους ιστορία* [Words have their own history]. https://sarantakos.wordpress.com/2013/11/.../mmm/ (checked 14/01/2017).

Sebeok, Thomas. 1974. *Structure and texture: Selected essays in Cheremis verbal art.* The Hague/Paris: Mouton.

Setatos, Michalis. 1994. Διλεκτικές και παραπλήσιες μονάδες της ΚΝΕ [Bilexic and similar units of SMG]. *«Γλωσσολογικές μελέτες». Επιστημονική επετηρίδα της φιλοσοφικής σχολής.* Περίοδος Β΄- Τεύχος Τμήματος Φιλολογίας Α.Π.Θ. Παράρτημα 4. 187–201

Sherzer, Joel. 1990. *Verbal art in San Blas. Kuna culture through its discourse.* Cambridge: Cambridge University Press.

Southern, Mark R. V. 2005. *Contagious couplings: Transmission of expressives in Yiddish echo phrases.* Westport, CT/London: Praeger.

Stolz, Thomas. 2008. Total reduplication vs. echo-word formation in language contact situations. In Peter Siemund & Noemi Kintana (eds.), *Language contact and contact languages,* 107–134. Amsterdam/Philadelphia: John Benjamins.

Stolz, Thomas. this volume. (Non-)canonical reduplication.

Stolz, Thomas, Cornelia Stroh & Aina Urdze. 2011. *Total reduplication. The areal linguistics of a potential universal.* Berlin: Akademie Verlag.

The National Herald GR. Greek version of the Greek-American Journal *Εθνικός Κήρυξ/ The National Herald*, No 31120, at http://herald1428.rssing.com/browser.php?indx=25197478&item=4392 (last access October 1st, 2016).

Thomopoulos, Ioannis. 1986. Το ύφος του νεοελληνικού λόγου [The style of Modern Greek speech]. *Νέα Εστία* 1414–1418. 729–737.

Triandafillidis, Manolis. 1938. *Νεοελληνική γραμματική. Ιστορική εισαγωγή*, ανατύπ. στο *Άπαντα Μ. Τριανταφυλλίδη*, τ. 3 [Modern Greek grammar. Historical introduction, reprint in M. Triantafyllidis collected writings]. Θεσσαλονίκη: ΙΝΣ, 1981.

UMGD 2014 = Χρηστικό λεξικό της νεοελληνικής γλώσσας [Utilitarian Modern Greek dictionary]. Athens: Academy of Athens.

Valavanis, Ioannis G. 1892. Ζώντα μνημεία της ανά τον Πόντον ιδιωτικής [Living monuments of the Pontic dialect]. In *Αρχεία της Νεωτέρας Ελληνικής Γλώσσης εκδιδόμενα υπό του Συλλόγου 'Κοραή', τ. Α΄, τεύχος γ΄ και δ΄ (Τμήμα Α΄ Λεξιλόγιον)*. Αθήναι.

Versnel, Henrik S. 2002. The poetics of the magical charm: An essay on the power of words. In Paul Allan Mirecki & Marvin W. Meyer (eds.), *Magic and ritual in the ancient world,* 105–159. Leiden/Boston/Köln: Brill.

Vounchev, Boris. 2016. On some properties of the affixoids in Modern Greek. Available online at http://ngf-bg.com/wordpress/wp-content/uploads/Vounchev_Affixoids_ENG-1.pdf.

Yip, Moira. 1992. Reduplication as alliteration and rhyme. *GLOT International* 4. 1–7.

Zwicky, Arnold M. & Geoffrey K. Pullum. 1987. Plain morphology and expressive morphology. In Jon Aske, Natasha Beery, Laura Michaelis & Hana Filip (eds.), *Proceedings of the Thirteenth Annual Meeting of the Berkeley Linguistics Society. General Session and Parasession on Grammar and Cognition*, 330–340. California: Berkeley Linguistics Society.

Julia Nintemann

Circumventing bisyllabic minimality

How to reduplicate monosyllabic verbs in (South-)East African Bantu languages

Abstract: This paper analyzes and compares how different East and South East African Bantu languages reduplicate monosyllabic verbs. Usually, Bantu languages require a minimally bisyllabic reduplicant in verb reduplication, which is why the reduplication of monosyllabic verbs is problematic and calls for avoidance strategies in order to circumvent the problem that arises in relation to the *bisyllabic minimality condition* (BSC).

Keywords: verb reduplication, monosyllabic verbs, bisyllabic minimality, augmentation, Bantu

1 Introduction

Reduplication is a common feature in Bantu languages and affects all kinds of word classes.

> In this, as in most aspects of Bantu grammar, the story is one of theme and variations: The reduplications in question show great similarity both in structure and in meaning, but also interesting differences. Nowhere are these differences more pronounced – or more significant – than in verb reduplication. (Hyman 2009: 178)

The reduplication of verbs certainly is the most complex and elaborate type of reduplication in Bantu languages. In comparison to the reduplication of other word classes, verb reduplication has a number of rules to abide to and shows an interesting set of characteristics as to the form of the reduplicant. Even the functions of the reduplication of verbs are more diverse than those of other word classes.[1] Reduplicated verbs in Bantu languages mostly express intensification, repetition, habitualness, diminution, distributiveness, or aimlessness, and sometimes have a pejorative meaning.

---

1 The reduplication of adjectives, for example, normally has only two functions: intensification, or on the contrary, diminution, e.g., Ekegusii *mambi~mambia* 'very early' (Mecha 2010: 57) or Swahili *dhaifu~dhaifu* 'weakish' (Brauner & Bantu 1967: 148).

Julia Nintemann: FB 10: Language Sciences, University of Bremen, Bibliothekstr. 1, 28359 Bremen, Germany. E-mail: jnintemann@uni-bremen.de

https://doi.org/10.1515/9783110599329-005

One of the most striking features of verb reduplication in Bantu languages is that they usually require a minimally bisyllabic reduplicant. Due to this *bisyllabic minimality condition* (BSC), monosyllabic verbs are problematic in that the verb stem itself is shorter than the required two syllables for the reduplicant. Although Bantu languages usually have only a small set of monosyllabic verbs, even these verbs have to be able to reduplicate somehow. Thus, there must be a way to circumvent the problem of bisyllabic minimality.

In this paper, the reduplication of monosyllabic verbs in twenty-four East and South-East African Bantu languages reaching from the equatorial regions of Kenya and Uganda to Swaziland and the North-Eastern part of South Africa will be analyzed (cf. Appendix).[2] How do the different languages compensate for the lack of a second syllable of the base? Can languages have more than one strategy to circumvent the problem of bisyllabic minimality? How do the different strategies distribute over the twenty-four languages of ten different countries? In the following sections, I will try to get to the bottom of these questions and give some deeper insights into the reduplication of monosyllabic verbs in the twenty-four languages under discussion.

For analyzing the forms of reduplication in the twenty-four sample languages, I take a functionalist point of view. Based on Mel'čuk (1996) and Stolz et al. (2011) the copy will be called *reduplicant*, while "the portion of a meaning-bearing unit ("signifiant") within which reduplication applies" (Stolz et al. 2011: 40) will be called *domain*. Finally, "the portion of a [domain] which is reduplicated" (Stolz et al. 2011: 40) will be the *base*.

In order to fully grasp how the reduplication of monosyllabic verbs differs from bisyllabic or polysyllabic verbs, a short overview of Bantu verb reduplication in general will be given in Section 2. In Section 3, I will have a closer look at the morphophonological structure of monosyllabic verbs in Bantu languages and in what way the BSC poses a problem for these verbs. An analysis of the different strategies used for circumventing the problem of bisyllabic minimality will be given in Section 4. Finally, Section 5 will evaluate and conclude the preceding analysis.

2 Note that the languages under discussion reflect only the existent 'doculects', a term introduced by Jeff Good that "refer[s] to the variety of the 'language' that ends up in documentation" (Bowern 2008: 8). The results may vary if actual field work is conducted.

2 How verbs are reduplicated in Bantu languages

Stolz et al. (2011: 8–70) discuss the prototype of reduplication and make the following assertion: "Be the prototype of reduplication a reduplicative construction which has the following properties. It is total, exact, contiguous and continuous" (Stolz et al. 2011: 42). Fulfilling four of the six parameters introduced by Mel'čuk (1996), the prototype has to be a complete and exact copy of the base, which is directly attached to either the left or the right of the base. By having a look at how Bantu verbs usually reduplicate, it becomes clear that the reduplication does not completely correspond to the prototype, as there are a lot of cases in which the reduplication is neither total nor exact. Hyman (2009) investigates how Bantu languages reduplicate verbs and finds some variation in the forms of verb reduplication.

> While some Bantu languages exhibit total reduplication of the stem constituent consisting of the verb root + suffixes, others place maximum size constraints on the reduplicant and/or disallow certain suffixes, e.g. inflectional endings, from appearing within it. (Hyman 2009: 178)

For a start, we have a look at the specifics of Bantu verb reduplication. A number of Bantu languages make use of a so-called default vowel, which must (or can) serve as the nucleus of the reduplicant's last syllable. Kikuyu verb reduplication perfectly illustrates the use of a default vowel. Consider the following examples:[3]

(1)		Kikuyu				[Peng 1991: 61]
	a.	*rora*	'see'	→	***rora***~*rora*	'see a little'
	b.	*cañka*	'pop'	→	***cara***~*cañka*	'pop a little'
	c.	*hũngũtũka*	'wander'	→	***hũnga***~*hũngũtũka*	'wander a little'

It is noticeable that the reduplicant is bisyllabic in all three cases, which is due to Kikuyu being a language in which the reduplicant has to be not only minimally but also maximally bisyllabic. Apart from that, it is striking that all of the reduplicants end with the vowel /*a*/. Peng (1993: 18) argues that "Kikuyu *a* appears to be a default vowel". The default-/a/ "consistently appears as the second syllable nucleus of the reduplicant" (Peng 1991: 61), no matter if this posi-

3 In this and all of the following examples, the reduplicant will be in boldface and separated from the base with a tilde. The examples will be given with glosses, apart from a few exceptions where it is not necessary for the illustration of the phenomenon.

tion is empty or filled (Peng 1991: 63). If the second syllable is filled, the default-/a/ overwrites every vowel other than /*a*/. Thus, the canonical reduplicant does always consist of CVC-/*a*/ or CV-/*a*/.

Downing (1999a) argues that /*a*/ is usually used as an unmarked *Inflectional Final Suffix* (IFS) for Bantu verbs and thus, the reduplicants "phonologically resemble a canonical Bantu verb stem, ending with the unmarked Inflectional Final Suffix" (Downing 1999a: 11). In Kikuyu the default-/a/ is obligatory and always forms the second syllable's nucleus of the obligatorily bisyllabic reduplicant. In Bukusu reduplication, for example, there are different rules for the use of the default-/a/, as the examples in (2) suggest.

(2)	Bukusu		[Downing 2004: 74]
a.	*xŭu-rem-er-an-a*	→	*xŭu-**rem-a**~rem-er-an-a*
	INF-cut-APPL-REC-FV		INF-cut-FV~cut-APPL-REC-FV
	'to cut for each other'		'to cut for each other frequently'
b.	*xŭu-rem-er-an-a*	→	*xŭu-**rem-er-a**~rem-er-an-a*
	INF-cut-APPL-REC-FV		INF-cut-APPL-FV~cut-APPL-REC-FV
	'to cut for each other'		'to cut for each other frequently'

As can be seen in (2), the default-/a/ does not necessarily form the second syllable's nucleus in Bukusu as it is the case in Kikuyu. While the prefix is normally not part of the reduplication, the suffixes may be copied along with the verb stem. The default-/a/ can either replace all suffixes (2a) or only part of them (2b). Furthermore, "a bisyllabic RED cannot be /a/-final if the result is not a subconstituent of the Base stem" (Downing 2004: 74).

(3)	Bukusu		[Downing 2004: 75]
a.	*xuu-[kelul]-a*	→	*xuu-**kelu**~kelul-a*(**xuu-**kela**~kelula*)
	INF-separate-FV		INF-separate~separate-FV
	'to separate; turn over'		'to separate frequently'
b.	*xuu-[kel-ul]-a*	→	*xuu-**kel-a**~kel-ul-a*
	INF-measure-REV-FV		INF-measure-FV~measure-REV-FV
	'to measure out'		'to measure out frequently'

As the examples in (3) show, it is not possible to have a bisyllabic reduplicant with final /*a*/ if the base is longer than CVC. Unlike in Kikuyu, the default-/a/ may not overwrite parts of the verb stem. Instead, the nuclear vowel of the domain's second syllable is the final vowel used for the reduplicant. Thus, although the reduplicant "is generally /*a*/-final, even if the [domain] contains another inflectional suffix" (Downing 2004: 74), there are instances, in which the default-/a/ is not used.

In Lusaamia, verbs reduplicated with the default-/a/ have a slightly different meaning than the ones reduplicated without, as seen in (4).

(4) Lusaamia [Marlo 2002: 21–23]

a. *a-**tem-aaŋg-a**~tem-aaŋg-a*
3SG-chop-HAB-FV~chop-HAB-FV
'He habitually chops in the same way and continues to chop over and over'

b. *a-**tem-a**~tem-aaŋg-a*
3SG-chop-FV~chop-HAB-FV
'He chops habitually, but the chopping takes place at random intervals'

While the reduplication including the habitual suffix plus final vowel (4a) implies doing something continuously, steadily and in regular intervals, the default-/a/ is used to express a more inconsequent way of doing something continually and in varying intervals (4b). Note that the final vowel /*a*/ in (4a) is the unmarked IFS succeeding the habitual suffix -*aaŋg* in both the base and the copy, while the default-/a/ in (4b) overwrites all other suffixes in the reduplicant. Similarly, derived verbs in Lusaamia might reduplicate in two different ways in order to produce either an iterative meaning (reduplicating the verb with all derivational suffixes) or a distributive meaning (/a/-final reduplicant) (Marlo 2004: 262).

The examples given for illustrating the use of the default-/a/ already show a variety of reduplicated forms. Formally, both total and partial reduplications occurred as well as both exact and non-exact. While all the reduplications of Bantu verbs seem to fulfill the requirement of being continuous, Downing (2004: 75) introduces a type of infixing reduplication for verbs with a "longer V-initial stem" in Bukusu (5a) as well as a type of reduplication in Bukusu in which "often it is the final two syllables of the Base which are repeated" (Downing 2004: 79) and for which she argues that the reduplicant is rather infixed than suffixed[4] (5b).

4 She argues that the reduplicant is rather infixed than suffixed because of two reasons:
1. [I]nflectional morphology is found on the second copy, not on the first [...]. The default IFS /a/ replaces the inflectional suffix in the first copy.
2. [T]he second copy does not always include just the final two syllables of the input base stem. Instead, both the second copy and the first can be longer than two syllables [...]. The only constraint on their length is that the first copy must be a subconstituent of the second copy.

(5) Bukusu [Downing 2004: 76]

a. *xú-xwa-aakam-a* 'to have come to an end' → *xú-xw-aa~**kama**~kama*

b. *ßa-kâcuule* 'they conversed' → *ßa-kâ~**cula**~cuule*

The examples in (5) show a type of reduplication which fails to fulfill the requirements of being total and continuous, as the reduplicant interrupts the segmental chain of the domain. Although this seems to be a rare exception, it shows that the reduplication of verbs in Bantu languages can be quite far from the prototype.

Nevertheless, there is a general guideline for the reduplication of verbs, which is followed in most cases:

- The reduplicant is attached to the left of the base (verb stem).
- The reduplicant has to be minimally bisyllabic.
- The default vowel /*a*/ must or may be used as the nucleus of the reduplicant's final syllable.
- Prefixes are not part of the reduplication.
- Suffixes may be reduplicated along with the verb stem.

Although only a small portion of the reduplicated verb forms in Bantu languages can probably be called prototypical in the way that Stolz et al. (2011) define the prototype of reduplication, there is a small number of forms which may be labeled the prototype of Bantu verb reduplication. However, as the one rule which absolutely has to be followed is that the reduplicant has to be minimally bisyllabic, monosyllabic verbs require some extra rules which distance them even from the prototype of Bantu verb reduplication.

3 Defining the monosyllabic verb

As Bantu languages usually have open syllables which are formed by adding an IFS, verb stems with a CVC structure like Kikuyu *ror* 'see' in (1a) do not belong to the category of monosyllabic verbs. In Proto-Bantu, "[t]he syllable structures allowed [...] were limited to [...] (a) CV, CVV (b) V, N" (Hyman 2003: 43) and "[m]ost Bantu languages maintain a close approximation of the [Proto-Bantu] situation as far as syllable structure is concerned" (Hyman 2003: 44). Verbs do

[A]ll of the derivational suffixes of the input verb stem occur on the second copy, while the first copy contains only a subset. (Downing 2004: 79)

not occur without verb extensions and/or the Final Vowel which ensure that the verb has open syllables only.

For Kikuyu, Peng (1993: 15) defines the canonical root "as consisting of a syllable and an additional consonant: [σ.C]". With the Final Vowel /a/ attached, this makes a bisyllabic verb with two open syllables instead of a monosyllabic verb. Thus, verb stems with a CVC structure do not belong to the category of monosyllabic verbs and do not pose a problem for the BSC. The problematic cases are shorter than the canonical root, which means that they consist of CV, V or C.

In Kikuyu, there are a few verbs having a CV structure. These verbs do not normally pose a problem for the BSC as they take the default vowel /a/ just like any other verb when reduplicated (6a). Some of these supposedly monosyllabic verbs trigger vowel length (6b) and the BSC is still satisfied by the attachment of the default-/a/. Note the exception of a verb consisting of only V in (6c), where there is no default-/a/, but the BSC is still satisfied by using a different vowel.[5]

(6) Kikuyu [Mugane 1997: 11]

a.	*ki-a*	'to prepare porridge/beer'	→	***ki-a***~*ki-a*	'to prepare porridge/beer a little more'
b.	*rĩ-a*	'to eat'	→	***rĩĩ-a***~*rĩĩ-a*	'to eat a little'
c.	*ũ-a*	'to peel'	→	***ũo***-*ũo*	'to peel a little more'

These kinds of supposedly monosyllabic verbs are completely unproblematic for the BSC, as these verb stems do not have /a/ as nucleus, their CV-/a/ structure makes a perfectly bisyllabic reduplicant. Thus, the problem of not being able to reduplicate with the default-/a/ normally only arises when the verb stem consists of only C, CC or Ca and these verbs will be the object of study in the following section.

4 The different approaches

As monosyllabic verbs cannot be reduplicated in the same way as bisyllabic and polysyllabic verbs reduplicate in Bantu languages, there is a need for avoidance strategies to circumvent the problem of bisyllabic minimality. There are indeed

5 Peng (1993) notices that Benson (1964) lists three exceptions to the BSC: "i) ha-hahek 'chat a little'; ii) na-nat 'push somewhat right on'; and iii) tɛ-tiyer 'be a little afraid'" Peng (1993: 16). He cannot, however, give an explanation as to why these forms reduplicate this way.

several possibilities which are employed in different languages. Some languages even make use of more than one strategy depending on conditions such as tense or mood.[6]

4.1 Augmentation through suffixes

Extending the verb by inserting a suffix is one of the strategies commonly used in the reduplication of monosyllabic verbs in Bantu languages. In Lusaamia, for example, there are some verbs existing of only a consonant which can be reduplicated "in the present tense by epenthesizing [*itʃ*]" (Marlo 2002: 15), cf. (7).

(7)		Lusaamia					[Marlo 2002: 16]
	a.	*oxu-tʃ-a*	→	*a-**tʃ-itʃ-a**~tʃ-itʃ-aanga*			
		INF-go-FV		3SG-go-AUG-FV~go-AUG-HAB			
		'to go'		'He's always going'			
	b.	*oxu-t-a*	→	*a-**t-etʃ-a**-t-etʃ-aang(a)*	*ano*	*esindu*	
		INF-put-FV		3SG-put-AUG-FV~put-AUG-HAB	here	something	
		'to put'		'He's always putting something here'			

In both (7a) and (7b) the monosyllabic verb stems *-tʃ-* 'go' and *-t-* 'put', respectively, are extended by the insertion of the augment *-itʃ*. Together with the Final Vowel *-a* these verbs are perfectly bisyllabic so that the BSC does not pose a problem for the reduplication of these verbs anymore. Note that the habitual suffix *-aanga* is not used to compensate for the lack of a second syllable of the base.

Other than inserting /*itʃ*/, "[i]t is also possible to form a disyllabic I-Stem that can be reduplicated by adding the VCV perfective suffix *-ire*" (Marlo 2002: 16). Naturally, this can be used only for verbs in perfect tense. Here, the verb stem is reduplicated together with the perfective suffix, as seen in (8).

(8)	Lusaamia			[Marlo 2002: 16]
	oxu-ly-a	→	*y-aa-**li-ire**~li-ire*	
	INF-eat-FV		3SG-TM-eat-PERF~eat-PERF	
	'to eat'		'He ate repeatedly (derogatory)'	

6 Although usually the BSC comes into effect, there are some Bantu languages where total reduplication (TR) und partial reduplication (PR) are distinguished in the grammars. In the case of PR, the reduplicant usually is monosyllabic. For Bemba, Kula (2002: 100) states that "most of the partial reduplication forms are lexical, and the process cannot be regarded to be as productive as total reduplication", e.g., *tun~tumba* 'carry a heavy load'.

In contrast to the habitual suffix -*aanga* in (7), the perfective suffix -*ire* is used to make up for the lack of a second syllable. Thus, there are restrictions as to which suffixes may be reduplicated.

In Kuria, a similar strategy is used to satisfy the BSC.

> Some tenses, such as the Immediate Past Anterior, have a perfective morpheme as a suffix. When monosyllabic verbs are used in such tenses, they are lengthened by the perfective morpheme and consequently they are able to reduplicate. (Mwita 2008: 237)

(9)	Kuria			[Mwita 2008: 238]
a.	*h-a*	→	*ßa-a-**h-er-e**~h-er-e*	
	give-FV		SM-TM-give-PERF-FV~give-PERF-FV	
	'give'		'they have just given repeatedly'	
b.	*sj-a*	→	*ßa-a-**se-er-e**~se-er-e*	
	grind-FV		SM-TM-grind-PERF-FV~grind-PERF-FV	
	'grind'		'they have just ground repeatedly'	

As the insertion of the perfective suffix is limited to tenses that express past events, this kind of extension of the monosyllabic verb stem cannot always be used in order to satisfy the BSC. In other cases, an object prefix can be added as will be discussed in Section 4.2.

In Ekegusii, monosyllabic verbs with glided vowels are extended by inserting suffixes which "compensate[...] for the deficiency of the root that blocks reduplication" (Mecha 2010: 34). Consider the examples in (10).

(10)	Ekegusii			[Mecha 2010: 34]
a.	*oko-ry-a*	→	*oko-**ri-e-gw**-a~ri-e-gw-a*	
	INF-eat-FV		INF-eat-PERF-by-FV~eat-PERF-by-FV	
	'to eat'		'to be eaten continually'	
b.	*ogo-tw-a*	→	*ogo-**tw-e-rw-a**~tw-e-rw-a*	
	INF-pluck-FV		INF-pluck-PERF-for-FV~pluck-PERF-for-FV	
	'to pluck'		'to be plucked for now and again'	

The material used for extending verb stem is infixed between the root and the Final Vowel /*a*/ as seen in (10). This process cannot take place without changing the tense as the suffixes constitute a combination of the "perfective {-e-} marking the present perfect tense in Ekegusii and the post-clitics -rw-,-gw- marking the post-positional meanings, "by" and "for" respectively" (Mecha 2010: 35). Furthermore, the suffixes can take two different phonological shapes, one keeping the suffixes' glides plus the final vowel /*a*/, the other replacing them with /*u*/ in the reduplicant, so that *ogotwerwa~twerwa* will become *ogotweru~twerwa* both having the meaning of 'to be plucked for now and again' (cf. Mecha 2010: 34).

Just like in the cases of Lusaamia and Kuria discussed and (8) and (9), respectively, the suffix that is used for extending the monosyllabic verb stem is a perfective suffix, so that the reduplication of the verb is accompanied by a change of tense. A different way of reduplicating certain monosyllabic verbs in Ekegusii will be discussed in Section 4.3.

Ciyao is another Bantu language in which a monosyllabic verb can be reduplicated if it is extended by a tense marking suffix. Mtenje (2003: 4) gives a number of examples of monosyllabic verbs which are reduplicated along with the past tense suffix *-ile* (or *-ele*), cf. (11).

(11)		Ciyao		[Mtenje 2003: 46]
	a.	*n-á-l-iíle*	→	*n-á-**l-iilé**~l-iile*
		1SG-TM-eat-PST.FV		1SG-TM-eat-PST.FV~eat-PST.FV
		'I ate'		'I ate frequently'
	b.	*w-á-mw-eéle*	→	*w-á-**mw-eelé**~mw-eele*
		3SG-TM-drink-PST.FV		3SG-TM-drink-PST.FV~drink-PST.FV
		's/he drank'		's/he drank frequently'

If the monosyllabic verb does not occur with a tense marking suffix, there is another way to circumvent the BSC in Ciyao, cf. Section 4.4.

Zulu behaves differently than the languages discussed above in that the past tense suffix *-il-* is not allowed to take part in the reduplication (Cook 2013: 58). Instead, an augment *-yi* or *-y* is used for the satisfaction of the BSC (Cook 2013: 5). This is illustrated in (12).

(12)		Zulu		[Cook 2013: 58]
	a.	*u-dl-il-e*	→	*u-**dl-a-yi**~dl-il-e*
		2SG-eat-PST-FV		2SG-eat-FV-AUG~eat-PST-FV
		'you ate'		'you did a bad job eating'
	b.		→	**u-**dl-il-a**~dl-il-e*
				2SG-eat-PST-FV~eat-PST-FV
				'you did a bad job eating'

The form **udlila~dlile* in (12b) where the past tense suffix takes part in the reduplication is not possible. The only acceptable way of reduplicating a monosyllabic verb in past tense is to make use of the augment *-yi*. As we will see in Section 4.2, there is another possibility of reduplicating monosyllabic verbs in Zulu. However, suffixes other than the augment are generally not allowed in the reduplication process.

Zulu is not the only language which makes use of the augment *-yi* for the extension of monosyllabic verbs. Swati and Ndebele are two further examples of Bantu languages in which "a dummy second syllable *-yi* [...] fills out the bisyllabic template" (Hyman 2009: 186).

(13) a. Swati [Peng 1991: 94]
n-a → ***n-a-yi***~*n-a*
rain-FV rain-FV-AUG~rain-FV
'rain' 'rain a little'
b. Ndebele [Hyman 2009: 185]
w-a → ***w-a-yi***~*w-a*
fall-FV fall-FV-AUG~fall-FV
'fall' 'fall here and there'

The augment *-yi* is used to make the reduplicant bisyllabic. Note that *-yi* is inserted after the default-/a/ in the examples in (13) as well as in (12a), although /a/ usually follows all other verbal extensions, e.g., Ndebele *theng-is-el-an-a* 'sell to each other' where the final vowel /a/ is preceded by the causative, applicative and reciprocal suffixes, respectively (Khumalo 2014: 144).

It is quite surprising that tense marking suffixes cannot be used in all of the languages where the reduplicant is expanded by inserting a suffix, as some of these suffixes may be part of the reduplication even with bisyllabic or polysyllabic verbs in some Bantu languages. Even if the domain has a tense marking suffix, it is replaced by a dummy syllable in the reduplicant as the Zulu example in (12) shows.

4.2 Augmentation through prefixes

In a paper about Bantu reduplication Hyman et al. (1999: 175) state that "[w]ith one major exception [...], prefixes are irrelevant to stem reduplication". This major exception is that in some languages, prefixes are part of the reduplication of monosyllabic verbs in order for the reduplicant to be bisyllabic.

In Bukusu, for example, the (infinitive) prefix, although normally not part of the reduplication, "must be reduplicated along with the stem" (Downing 2004: 75), cf. (14).

(14) Bukusu [Downing 2004: 75]
a. *xúu-ly-a* → *xúu-ly-a*~***xuu-ly-a***
INF-eat-FV INF-eat-FV~INF-eat-FV
'to eat' 'to eat repeatedly'

b.	*xúu-a*	→	*xúu-a~**xuu-a***
	INF-give		INF-give~INF-give
	'to give'		'to give repeatedly'

With the insertion of the infinitive prefix, a monosyllabic verb is able to reduplicate. However, there is an alternative for monosyllabic verbs which is more common than the reduplicative pattern. In order to express the same content as the reduplication, i.e., "a meaning like 'repeatedly' or 'carelessly'" (Downing 2004: 73), a suffix can be added to the stem. Instead of a reduplicated form *xúulya~xuulya* 'to eat repeatedly', there is a non-reduplicative alternative *xúu-líichaka* 'to eat repeatedly', in which the *-ich-ak-* suffixes are used.

Like Bukusu, Kinyakyusa uses the prefixes of monosyllabic verbs for the satisfaction of the BSC of the reduplicant, cf. (15).

(15)	Kinyakyusa		[Lusekelo 2009: 21]
	ku-nw-a	→	*ku-nw-a~**ku-nw-a***
	INF-drink-FV		INF-drink-FV~INF-drink-FV
	'to drink'		'to drink recklessly

As Lusekelo (2009: 21) states, "the infinitive marker -ku- is copied in monosyllabic verbs". Both Downing (Bukusu) and Lusekelo (Kinyakyusa) only give examples of reduplication of monosyllabic verbs in their infinitive form and without any other prefixal material. The examples of Swahili in (16) show, that it is not always the infinitive prefix which is copied for the satisfaction of the BSC.

(16)		Swahili		[Park 1995: 298]
	a.	*ku-l-a*	→	*ku-l-a~**ku-l-a***
		INF-eat-FV		INF-eat-FV~INF-eat-FV
		'to eat'		'to eat repeatedly'
	b.	*ku-ki-l-a*	→	*ku-ki-l-a~**ki-l-a***
		INF-OBJ.3SG-eat-FV		INF-OBJ.3SG-eat-FV~OBJ.3SG-eat-FV
		'to eat it'		'to eat it repeatedly'

In order to have a bisyllabic reduplicant, "material before the stem is also copied for monosyllabic stems" (Park 1995: 299). In (16a), the infinitive prefix is copied along with the monosyllabic verb stem. In (16b), the monosyllabic verb has an object prefix between the infinitive prefix and the verb stem. In this case, the prefix closest to the verb stem, i.e., the object prefix, is part of the reduplication, while the prefix further away, i.e., the infinitive prefix, is not reduplicated. For lack of uncontroversial data, I am not yet in a position to confirm if this is also the case in Bukusu and Kinyakyusa, respectively, but it is conceivable that the insertion of the infinitive prefix is not the only possibility for satisfying the BSC.

It is clear, however, that Swahili is not the only Bantu language, in which the prefix closest to the verb stem is copied to make the reduplicant bisyllabic, cf. (17).

(17) Chindamba [Edelsten & Lijongwa 2010: 93]

a. *a-ku-ly-a* → *a-ku-ly-a~**ku-ly-a***
SBJ.3SG-TM-eat-FV SBJ.3SG-TM-eat-FV~TM-eat-FV
'he eats' 'he is continuously eating'

b. *a-ku-va-p-a* → *a-ku-va-p-a~**va-p-a***
SBJ.3SG-TM-OBJ.3PL-give-FV SBJ.3SG-TM-OBJ.3PL-give-FV~OBJ.3PL-give-FV
'he gives them' 'he is continuously giving them'

Just like in Swahili, in Chindamba "the preceding syllable is copied by the reduplication regardless of its function" (Edelsten & Lijongwa 2010: 93). Thus, the tense marker *-ku-* is copied in (17a), while in (17b), the object marker *-va-*, which is closer to the verb stem than the tense marker, is a part of the reduplication.

In some languages, however, there is a restriction as to which prefixes may be copied. In Kitharaka, the object marker and the reflexive prefix "can occur in the reduplicant if the root is made up of only a consonant" (Muriungi 2008: 110). Consider the examples in (18).

(18) Kitharaka [Muriungi 2008: 110]

a. *î-p-a* → *î-pa~**î-p-e***
REFL-give-FV REFL-give-FV~REFL-give-FV
'give yourself' 'give yourself quickly'

b. *mû-p-a* → *mû-p-a~**mû-p-e***
OBJ.3SG-give-FV OBJ.3SG-give-FV~OBJ.3SG-give-FV
'give him/her' 'give him/her quickly'

While these prefixes may be reduplicated with monosyllabic verbs, others may not. Muriungi (2008: 109–110) shows that tense markers, negation or subject agreement prefixes cannot be reduplicated. Thus, a form **ba-gû-p-a~gû-p-a* 'they have given quickly' with *ba-* as subject agreement and *-gû-* as a tense marker is not possible. This means that monosyllabic verbs can be reduplicated only in quite a limited context.

In Kuria, where it is possible to insert a perfective suffix to satisfy the BSC (cf. Section 4.1), there is also the possibility to prefix an object marker, which is then reduplicated along with the monosyllabic verb stem. This "is particularly seen in tenses that do not include the perfective suffix" (Mwita 2008: 238).

(19) Kuria [Mwita 2008: 238–239]
ta-h-a → *ta-mo-h-a~**mo-h-a***
TM-give-FV → TM-OBJ.3SG-give-FV~OBJ.3SG-give-FV
'(do) give' → '(do) give him repeatedly'

In (19), the monosyllabic verb *h-a* 'give' in the hortatory imperative can be reduplicated with the insertion of the object marker *-mo-*.

As we have seen in Section 4.1, a monosyllabic verb in Ndebele is usually able to reduplicate by adding the so-called dummy syllable *-yi*. However, if the monosyllabic verb stem is preceded by a class prefix, this class prefix may be copied along with the verb stem, cf. (20).

(20) Ndebele [Hyman 2009: 191]
úku-zi-dl-a → *úku-zi-dl-a~**zi-dl-a***
INF-OBJ.CL10-eat-FV → INF-OBJ.CL10-eat-FV~OBJ.CL10-eat-FV
'to eat them' → 'to eat them here and there'

The class prefix may only be copied with monosyllabic verb stems. It is not possible for them to be part of the reduplication when they occur with longer verb stems. It is, however, not obligatory to reduplicate monosyllabic verbs with the class prefix, even if there is one. The very same verb can also be reduplicated with the augment *-yi*. Thus, both forms *úku-zidla~zidla* and *úku-zi-dlayi~dla* are possible and employ the same meaning 'to eat them here and there'.

Zulu is very similar in that there is an alternative to the reduplication with the augment *-yi*. A monosyllabic verb may reduplicate with the prefixal long form present marker *-ya-*, as example (21) shows.

(21) Zulu [Cook 2013: 58]
u-ya-dl-a → *u-ya-dl-a~**ya-dl-a***
2SG-PRS-eat-FV → 2SG-PRS-eat-FV~PRS-eat-FV
'you eat' → 'you do a bad job eating'

Again, there are two ways of repeating the same verb. Thus, the form *u-ya-dlayi~dla* 'you do a bad job eating' is perfectly acceptable as well.

In Kibondei, "monosyllabic stems will copy prefixal material into the reduplicant to satisfy a binarity constraint that the reduplicant be at least bisyllabic" (Lee 2013: 138). In contrast to the examples above, it is not always the whole CV-structured prefix that is copied, but the CV structure may be broken apart so that only V will be copied into the reduplicant, cf. (22).

(22) Kibondei [Lee 2013: 139]
a. *ka-d-a* → *ka-d-a~**a-d-a***
3SG.PST-eat-FV → 3SG.PST-eat-FV~AUG-eat-FV
's/he ate' → 's/he was eating'

b.	*á-dí-d-a*	→	*á-dí-d-a~**dí-d-a***
	3SG.PRS-OBJ.3SG-eat-FV		3SG.PRS-OBJ.3SG-eat-FV~OBJ.3SG-eat-FV
	's/he eats it'		's/he is eating it'

Due to phonological reasons, the prefix *ka-* in (22a) is broken apart, while *-dí-* in (22b) retains its CV structure (cf. Lee 2013: 139–140).

The reduplication of monosyllabic verbs in Makonde is similar to the Kibondei case in that "the reduplicated part includes the copy of the preceding vowel" (Kraal 2005: 48).

(23)	Makonde		[Kraal 2005: 48]
a.	*kú-ly-á*	→	*kú-ly-á~**ú-ly-á***
	INF-eat-FV		INF-eat-FV~AUG-eat-FV
	'to eat'		'to keep on eating'
b.	*và-lì-ly-á*	→	*và-lì-ly-á~**ì-ly-à***
	3PL-OBJ.3SG-eat-FV		3PL-OBJ.3SG-eat-FV~AUG-eat-FV
	'they eat it'		'they keep on eating it'

As the verb stem in (23a) is preceded by the infinitive prefix *ku-*, the reduplicant on the right includes /*u*/. In (23b), on the other hand, the verb stem is preceded by the object marker *-li-*, which is why the reduplicant has /*i*/ as the initial vowel. Depending on the preceding syllable, the second copy may come with different initial vowels.

What strikes the eye when looking at the reduplicated monosyllabic verbs with an inserted prefix is that the reduplicant seems to be attached to the right rather than to the left of the verb stem. If we compare the two forms of the reduplicated verb 'eat' (present tense) in Zulu, we see that in the case of *u-ya-dlayi~dla* 'you do a bad job eating' it is obviously the first constituent that is the reduplicant as only the reduplicant is subject to the BSC. In the case of *u-yadla~yadla* 'you do a bad job eating', however, the two constituents involving the verb stem *-dl-* 'eat' look exactly the same. At first sight it seems that there is no way of distinguishing the reduplicant and the base. One might argue, that a form **u[yadla-dla]* might suffice if the left constituent was the reduplicant. However, as the long form present marker *-ya-* is already part of the base form *u-ya-dla* 'you eat' and prefixes are normally not part of the reduplication, it is not sufficient to have *-ya-* in the first constituent. Instead, the second constituent has to have the present marker *-ya-* as well, which might be a hint that the right constituent is the reduplicant here. That this is indeed the case becomes clear in languages like Kibondei or Makonde, in which the inserted prefix differs from the prefix preceding the first constituent in that it is usually only the vowel which makes it into the second constituent (cf. [22–23]). Thus, the reduplication of monosyllabic verbs which involves the augmentation through prefixes is non-prototypical for Bantu reduplication in that prefixes are copied along with the verb stem and that the reduplicant is attached to the right rather than to the left.

4.3 Bimoraicity

Ekegusii is one of the languages in which the reduplicant is not necessarily bisyllabic. Mecha (2010: 30–31) argues that "the base must be binary [at] some level for reduplication to occur" and that "bimoraicity compensates for disyllabicity, hence blocking the necessity of taking recourse to epenthesis". While monosyllabic verbs with glided vowels use the insertion of a suffix, others settle for a phonological lengthening of the vowel. Mecha (2010: 31) explains that the number of monosyllabic verbs in Ekegusii is so limited that this kind of reduplication occurs with only two verbs, which can be seen in (24).

(24)	Ekegusii			[Mecha 2010: 31]
a.	*ogo-t-a*	→	*ogo-**t-aa**~t-aa*	
	INF-pour-FV		INF-pour-FV[+L]~pour-FV[+L]	
	'to pour out'		'to pour out continually'	
b.	*ogo-s-a*	→	*ogo-**s-aa**~s-aa*	
	INF~diarrhoea-FV		INF-diarrhoea-FV[+L]~diarrhoea-FV[+L]	
	'to diarrhea'		'to diarrhea continually'	

In the Ekegusii examples in (24), the lengthening of the final vowel /*a*/ applies to both copies. In some languages like Kikerewe and Kimatuumbi, only the first constituent, i.e., the reduplicant has to be bimoraic (cf. [25a–b]). Other languages, e.g., Luganda, have a similarly bimoraic structure in both copies (cf. [25c]). Kinyarwanda in (25d) is another special case which will be discussed below.

(25) a.	Kikerewe			[Odden 1996a: 130]
	ku-gw-a	→	*ku-**gw-aa**~gw-a*	
	INF-fall-FV		INF-fall-FV[+L]~fall-FV	
	'to fall'		'to fall about'	
b.	Kimatuumbi			[Odden 1996b: 162]
	ly-á	→	***ly-áa**~ly-a*	
	eat-FV		eat-FV[+L]~eat-FV	
	'eat'		'eat frequently'	

c. Luganda [Hyman & Katamba 2006: 185]
ku-mw-a → *ku-**mw-aa**~mw-aa*
INF-shave-FV INF-shave-FV[+L]~shave-FV[+L]
'to shave' 'to shave a bit'

d. Kinyarwanda [Hyman 2009: 185]
gw-a → ***gw-aa**~gw-aan-a*
fall-FV fall-FV[+L]~fall-AUG-FV
'fall' 'fall here and there'

While the bimoraicity of the reduplicant alone (and possibly of the base as well) compensates for bisyllabicity in (25a-c), a form **gwaa-gwa* 'fall here and there' is not possible in Kinyarwanda. The monosyllabic verb stem can only reduplicate when "the base has been augmented by what looks like the reciprocal extension *-an-*"[7] (Hyman 2009: 186). Nevertheless, the same bimoraic structure of the first copy as in (25a–c) can be seen in (25d).

Although the reduplicant usually has to be minimally bisyllabic in most ([South-]East African) Bantu languages, there are a number of languages in which the reduplicant is actually monosyllabic. However, the reduplicant has to at least have a bimoraic structure, so that some kind of binary constraint is satisfied.

4.4 Multiple copying

Another commonly used method for satisfying the BSC is multiple copying. In this option the verb stem is copied twice instead of only once, as the examples in (26) show.

(26) a. Kinande [Downing 1999b: 64]
-swa → ***-swa-swa**~swa*
'grind' 'grind repeatedly'

b. Sukuma [Hyman 2009: 185]
-gwa → ***-gwa-gwa**~gwa*
'fall' 'fall here and there'

c. Lamba [Lodhi 2002: 8]
-ya → ***-ya-ya**~ya*
'go' 'go on and on and on'

7 As the suffix *-aan-* does not fulfill any kind of reciprocal purpose, it is glossed as augment in (25d).

d.	Ciyao			[Mtenje 2003: 46]
	-lyá	→	***-lyá-lya***~*lya*[8]	
	'eat'		'eat frequently'	
e.	Bemba			[Kula 2002: 100]
	-lya	→	***-lya-lya***~*lya*	
	'eat'		'eat carelessly'	
f.	Nyamwezi			[Schadeberg 2003: 89]
	-lya	→	***-lyaá-lyaa***~*lya*	
	'eat'		'eat frequently'	
g.	Cilungu			[Bickmore 2007: 318]
	-lwá	→	***-lwáa-lwáa***~*lwá*	
	'fight'		'fight repeatedly'	

As the examples (26a–g) show, there is not much variation among those languages that employ multiple copying for the satisfaction of the BSC. In (26a–e), the monosyllabic stem is copied twice without any phonological changes, while the reduplication of monosyllabic stems in (26f–g), i.e., in Nyamwezi and Cilungu, triggers vowel length in the reduplicant. Note that the vowel is lengthened in both copies that make up the reduplicant.

In contrast to augmenting the reduplicant with a prefix or suffix, multiple copying does not involve material that is normally not part of the reduplication. In Ciyao, multiple copying is only employed if there is no tense marking suffix (compare [11] and [26d]). As suffixal material may be copied in some Bantu languages, it seems that choosing the suffixes to extend the reduplicant for it to be bisyllabic is preferable in Ciyao as it is closer to the prototype of Bantu reduplication. I did not find any evidence of a language preferring the insertion of a prefix over the triplication of the verb stem. On the contrary, for Cilungu, Bickmore (2007: 318) gives an example *tú-kú-sháá-sháá~shá* 'we are grinding repeatedly' that shows that even though there are prefixes, namely the 1st person plural prefix *tú-* and the present progressive marker *kú-*, multiple copying is still preferred.

8 Mtenje (2003: 46) gives account of "a less widely spoken dialect [of Ciyao] which reduplicates monosyllabic verbs by copying the stem once only". Thus, instead of a form *ku-lyá-lya~lya* 'to eat frequently' (cf. [26d]), a form *ku-lyá~lya* 'to eat frequently' where the reduplicant consists of only one syllable and thus contravenes the BSC is used.

5 Conclusions

In the previous sections, four different ways of circumventing the problem of the BSC were presented. Except for the bimoraicity option, all of them are used to obtain a bisyllabic reduplicant. It was shown that languages can have more than one option to reduplicate monosyllabic verbs. Table 1 shows the distribution of the different options among the 24 languages under discussion.

Table 1: Distribution of options for satisfying the BSC.[9]

	Total number of languages	Share of languages
Augmentation through prefixes	10	41.7 %
Augmentation through suffixes	7	29.2 %
Multiple copying	7	29.2 %
Bimoraicity	5	20.8 %

Table 1 shows that the most commonly used option for the satisfaction of the BSC is the augmentation through prefixes. Ten out of twenty-four languages, i.e., 41.7 % of the languages under discussion use prefixes for the reduplication of monosyllabic verbs. While the augmentation through suffixes and multiple copying are used by almost 30 % of the languages each, bimoraicity is the least used option with slightly more than 20 %. Although there appears to be a tendency towards the augmentation through prefixes being most commonly used, the different options are spread relatively evenly among the sample languages. None of the options has a significantly higher or lower occurrence than the others.

However, although the augmentation through suffixes and multiple copying share the same numbers, it is noticeable that 100 % of the languages that make use of more than one option have the augmentation through suffixes as one of these options. In fact, just two languages (Lusaamia and Swati) reduplicate monosyllabic verbs only by involving suffixes. Thus, there appears to be a tendency towards inserting suffixes only under certain circumstances. This is, for example, the case in Ekegusii where the reduplicant is expanded by inserting a suffix only when the monosyllabic verb stem has a glided vowel (cf. [10]). If the respective requirements are not fulfilled, a different option is used. Thus,

9 Note that the percentages add up to more than 100 % because some languages make use of more than one option.

the augmentation through suffixes appears to have a more unsteady position than the other options.

If the numbers of the two augmentation strategies are added up, however, it becomes clear that augmenting the stem by inserting either prefixes or suffixes statistically dominates, as 70 % of the sample languages are affected. Thus, augmenting the monosyllabic stem with additional morphological material is preferred over bimoraicity and multiple copying.

If there is any significance at all in looking at an areal distribution with this small set of sample languages, one might perceive an accumulation of languages that use the augmentation through prefixes along the East coast of Africa, while multiple copying seems to be more frequent among the languages closer to the interior of the continent. Bimoraicity seems to be more prominent among the North Eastern Bantu languages. For the augmentation through suffixes, no areal clusters can be identified, which might again be a hint for this option to have a more unsteady position, or on the contrary, to be so popular that it is not restricted to a specific area.

As for the genetic distribution, there seem to be no clear patterns. Four of the five languages which employ bimoraicity belong to the Northeast Savanna Bantu phylum, while only one of them (Kimatuumbi) belongs to the Rufijic Bantu languages. However, as the sample is quite small and half of the languages under discussion belong to the Northeast Savanna Bantu phylum, which is the largest East Bantu microphylum, there is probably no significance to these numbers. As far as these sample languages are concerned, areal distribution appears to be a more decisive factor than genetic relatedness when it comes to the different options for satisfying the BSC.

Another issue that comes up when talking about a minimally bisyllabic reduplicant in the reduplication of monosyllabic verbs is the assignment to the commonly used categories of total and partial reduplication. While it is obvious that the reduplicated monosyllabic verbs do not qualify for PR, there are also some discrepancies when assigning them to the category of TR. Rubino (2005: 11) defines full reduplication (=TR) as "the repetition of an entire word, word stem (root with one or more affixes), or root". In the reduplication of monosyllabic verbs in Bantu languages, however, there are cases in which even more than the word stem is copied and the reduplicant is longer than the base. It is questionable if the term total reduplication fully grasps this form of reduplication. Although it is true that the stem is copied completely, neither the extension of the monosyllabic stem with prefixes or suffixes or at least vowel lengthening nor multiple copying are normally part of TR. The term total reduplication in contrast to partial reduplication implies a complete copy of the base rather than copying only a part of the base. However, the involvement of more material than the base offers is not in the definition of total reduplication. Taking into consideration also the rules for reduplicating bisyllabic and polysyllabic verbs, it may be more convenient to talk about (minimally) bisyllabic reduplication in Bantu languages instead of dividing each and every form of reduplication into only two categories.[10]

The issue of bisyllabic minimality in reduplication is not restricted to Bantu languages or the African continent. A number of Australian languages follow a two syllable structure when it comes to the size of the reduplicant (Fabricius 1998: 27). While "[i]n many languages with monosyllabic roots [...], the pattern of reduplication will change for these forms, and monosyllabic reduplication will apply" (Fabricius 1998: 27), in Ngiyambaa, monosyllabic verbs will only reduplicate when transitive roots become derived intransitives and thus become polysyllabic, so that the reduplicant can have a two syllable structure, cf. (27).

(27) Ngiyambaa [Fabricius 1998: 27]
ŋa:gi~*ŋa:-gi-ɟili-ɲa=na*
RED~look-REFL-PRS=3ABS
'She's more-or-less looking at herself'

As Hamidzadeh & Russell (2015: 98) explain, in Guaraní "additional material from the neighboring morphemes needs to be incorporated" for the reduplication of monosyllabic verbs. Usually, prefixal material is used to make the reduplicant minimally bisyllabic (28a). Sometimes, stressable suffixes are used for the satisfaction of the BSC (28b).

(28) Guaraní [Hamidzadeh & Russell 2015: 99, 103]
a. *a-po~**apo***
A1S-jump~RED
'I jump over and over'
b. *ha-'u-ka~**'uka***
A1S-eat-CAUS~RED
'I cause to eat over and over'

10 The absolute division into TR and PR is not only problematic for the monosyllabic verbs in Bantu languages. Kikuyu, for example, always follows the same rules for verb reduplication independently from the length of the stem; the reduplicant is always bisyllabic. As the reduplication always performs the same function, no matter if a bisyllabic verb (=TR) or trisyllabic verb (=PR) is reduplicated, it does not seem to make much sense to consider these forms to be categorically distinct. Again, it would be more convenient to talk about bisyllabic reduplication. For more information on this topic, see Nintemann (2016).

Other Tupi-Guaraní languages use similar strategies to extend the reduplicant in bisyllabic reduplication (cf. Rose 2005: 359). As there seems to be "a very neat dichotomy between the semantics of monosyllabic and disyllabic reduplications [...] irrespective of the length of the reduplicated stem" (Rose 2005: 352), a division into TR and PR would not make much sense. Instead, Rose uses monosyllabic and disyllabic reduplication to describe the two semantically different forms. Even though there is no such distinction in Bantu languages, which would emphasize the importance of considering not only the phonological shape of the reduplication but also its semantics, it would still be much more coherent to classify Bantu verb reduplication as (minimally) bisyllabic reduplication instead of distinguishing PR and TR, which do not even fully grasp the forms of reduplicated monosyllabic verbs.

The reduplication of monosyllabic verbs in languages in which the reduplicant is subject to bisyllabic minimality shows that PR and TR are not always the adequate benchmark for analyzing reduplication patterns. Thus, instead of trying to classify each and every form of reduplication as either PR or TR, more attention should be paid to the principles by which the reduplication is governed as well as to the functions the reduplications perform in the individual languages. If the reduplicant in a specific language is subject to bisyllabic minimality (and possibly also maximality) and the reduplication always performs the same function, there should not be a classification of the reduplication of monosyllabic, bisyllabic or polysyllabic bases as TR or PR, respectively, but a category which covers all of the forms, e.g., (minimally) bisyllabic reduplication.

The bisyllabic minimality condition in Bantu languages raises some interesting issues when it comes to the reduplication of monosyllabic verbs. As the base does not meet the requirements for the reduplicant, more phonological material has to be added in order to satisfy the BSC. The twenty-four languages of this sample showed four different solutions to this problem, all of them disagreeing more or less with the usual rules of Bantu verb reduplication. While analyzing the different strategies for satisfying the BSC, it must not be forgotten that it is exactly these forms of reduplicated monosyllabic verbs that show that there is a bisyllabic minimality in the first place.

Acknowledgments: This paper is based on one of the subsections of my Master thesis which treats reduplication in Kenyan languages (Nintemann 2016: 15–21). To gain more insights into the topic of the reduplication of monosyllabic verbs and the different options languages have to circumvent the problem of bisyllabic minimality, I have expanded the sample by nineteen languages spoken in a variety of countries in Africa. I am grateful to my supervisor Prof. Dr. Thomas Stolz who guided me through my studies and induced me to work on reduplication, as well as Dr. Christel Stolz who took up the task of being the second examiner for my Master thesis. Dr. Hitomi Otsuka and Maja Robbers deserve to be mentioned for the support they gave me by discussing my data. I also want to thank Dr. Aina Urdze for the organization of the workshop on non-prototypical reduplication and for encouraging me to give a talk and write this paper. Although I received a lot of help and support, the sole responsibility for form and content of this article lies with me.

Abbreviations

1, 2, 3	first, second, third person
A1S	first person singular Set-A person prefix
ABS	absolutive case
APPL	applicative
AUG	augment
BSC	bisyllabic minimality condition
C	consonant
CAUS	causative
CL	class marker
FV	final vowel
HAB	habitual
IFS	inflectional final suffix
INF	infinitive
L	long
N	syllabic nasal
OBJ	object
PERF	perfect
PR	partial reduplication
PRS	present tense
PST	past tense
REC	reciprocal
RED	reduplicant
REFL	reflexive
REV	reversive
SBJ	subject
SG	singular
SM	subject marker
TM	tense marker
TR	total reduplication
V	vowel

References

Benson, T. G. 1964. *Kikuyu-English dictionary.* Oxford: Clarendon Press.

Bickmore, Lee. 2007. *Cilungu phonology.* Stanford: CSLI Publications.

Bowern, Claire. 2008. *Linguistic fieldwork.* New York: Palgrave Macmillan.

Brauner, Siegmund & Joseph K. Bantu. 1967. *Lehrbuch des Swahili.* Leipzig: VEB Verlag Enzyklopädie.

Cook, Toni. 2013. Morphological and phonological structure in Zulu reduplication. University of Pennsylvania. PhD Thesis.

Downing, Laura. 1999a. Morphological constrains on Bantu reduplication. *Linguistic Analysis* 29(1/2). 6–46.

Downing, Laura. 1999b. Verbal reduplication in three Bantu languages. In René Kager, Harry van der Hulst & Wim Zonneveld (eds.), *The prosody-morphology interface*, 62–89. Cambridge: Cambridge University Press.

Downing, Laura. 2004. Bukusu reduplication. In Chege Githiora, Heather Littlefield & Victor Manfredi (eds.), *Trends in African linguistics* 5, 73–84. Lawrenceville NJ: Africa World Press.

Edelsten, Peter & Chiku Lijongwa. 2010. *A grammatical sketch of Chindamba – a Bantu language (G52) of Tanzania.* Köln: Rüdiger Köppe.

Fabricius, Anne H. 1998. *A comparative survey of reduplication in Australian languages.* München: LINCOM Europa.

Hamidzadeh, Khashayar & Kevin Russell. 2015. The phonology of reduplication in Paraguayan Guaraní. In Natalie Weber & Sihwei Chen (eds.), *Proceedings of the Eighteenth and Nineteenth Workshop on Structure and Constituency in the Languages of the Americas: Papers for WSCLA 18 & 19*, 95–108. Vancouver: UBC Working Papers in Linguistics.

Hyman, Larry M. 2003. Segmental phonology. In Derek Nurse & Gérard Philippson (eds.), *The Bantu languages*, 42–58. London: Routledge.

Hyman, Larry M. 2009. The natural history of verb-stem reduplication in Bantu. *Morphology* 19(2). 177–206.

Hyman, Larry M., Sharon Inkelas & Galen Sibanda. 1999. *Morphosyntactic correspondence in Bantu reduplication.* Berkely: Ms. University of California.

Hyman, Larry M. & Francis X. Katamba. 2006. The word in Luganda. In F. K. Erhard Voeltz (ed.), *Typological studies in language*, 171–193. Amsterdam: John Benjamins.

Khumalo, Langa. 2014. On the reciprocal in Ndebele. *Nordic Journal of African Studies* 23(3). 140–161.

Kraal, Pieter J. 2005. A grammar of Makonde (Chinnima, Tanzania). Universiteit Leiden. PhD Thesis.

Kula, Nancy C. 2002. The phonology of verbal derivation in Bemba. Universiteit Leiden. PhD Thesis.

Lee, Kent. 2013. Tone spread and reduplication in Kibondei. *The Journal of Studies in Language* 29(1). 129–150.

Lodhi, Abdulaziz Y. 2002. Verbal extensions in Bantu (the case of Swahili and Nyamwezi). *Africa & Asia* 2. 4–26.

Lusekelo, Amani. 2009. A description of Kinyakyusa reduplication. *SKASE Journal of Theoretical Linguistics* 6(2). 18–37.

Marlo, Michael R. 2002. *Reduplication in Luusamia.* Indiana University Working Papers in Linguistics Online 2.
Marlo, Michael R. 2004. Prefixing reduplication in Lusaamia: Evidence from morphology. In Akinbiyi Akinlabi & Oluseye Adesola (eds.), *Proceedings, 4th World Congress of African Linguistics, New Brunswick 2003*, 259–269. Köln: Rüdiger Köppe.
Mecha, Evans. 2010. *The phonology and morphology of Ekegusii reduplication: An optimality theoretic approach.* Saarbrücken: VDM Verlag Dr. Müller.
Mel'čuk, Igor Aleksandrovic. 1996. *Cours de morphologie générale.* Vol. 3: Troisième partie: *Moyens morphologiques.* Quatrième partie: *Syntactiques morphologiques.* Montreal: Presses de l'Université de Montréal.
Mtenje, Al. 2003. An optimality theoretic account of Ciyao verbal reduplication. In John Mugane (ed.), *Linguistic typology and representation of African languages*, 43–68. Trenton NJ: Africa World Press.
Mugane, John. M. 1997. *A paradigmatic grammar of Gĩkũyu.* Stanford: CSLI Publications.
Muriungi, Peter K. 2008. Phrasal movement inside Bantu verbs: Deriving affix scope and order in Kĩĩtharaka. University of Tromsø. PhD Thesis.
Mwita, Leonard C. 2008. Verbal tone in Kuria. University of California. PhD Thesis.
Nintemann, Julia. 2016. A comparative analysis of forms and functions of reduplication in Kenyan languages. In Christel Stolz & Thomas Stolz (eds.), *From Africa via the Americas to Iceland: Studies on reduplication and experiencers*, 1–93. Bochum: Brockmeyer.
Odden, David. 1996a. Patterns of reduplication in Kikerewe. *OSU Working Papers in Linguistics* 48. 111–149.
Odden, David. 1996b. *The phonology and morphology of Kimatuumbi.* Oxford: Clarendon Press.
Park, Jae-Ick. 1995. Minimality effects in Swahili. In Akinbiyi Akinlabi (ed.), *Theoretical approaches to African linguistics*, 295–312. Trenton NJ: Africa World Press.
Peng, Long. 1991. Swati and Kikuyu reduplication: Evidence against exhaustive copy. *Studies in African Linguistics* 22(1). 45–72.
Peng, Long. 1993. The canonical verb root and Kikuyu reduplication. *Journal of West African Languages* 23(2). 15–26.
Rose, Françoise. 2005. Reduplication in Tupi-Guarani languages: Going into opposite directions. In Bernhard Hurch (ed.), *Studies on reduplication*, 351–368. Berlin/Boston: Mouton de Gruyter.
Rubino, Carl. 2005. Reduplication: Form, function and distribution. In Bernhard Hurch (ed.), *Studies on reduplication*, 11–29. Berlin/Boston: Mouton de Gruyter.
Schadeberg, Thilo C. 2003. Derivation. In Derek Nurse & Gérard Philippson (eds.), *The Bantu languages*, 71–89. London: Routledge.
Stolz, Thomas, Cornelia Stroh & Aina Urdze. 2011. *Total reduplication: The areal linguistics of a potential universal.* Berlin: Akademie Verlag.

Appendix: List of sample languages

Bemba (Zambia)
Bukusu (Kenya)
Chindamba (Tanzania)
Cilungu (Tanzania)
Ciyao (Malawi)
Ekegusii (Kenya)
Kibondei (Tanzania)
Kikerewe (Tanzania)
Kimatuumbi (Tanzania)
Kinande (Republic of Congo)
Kinyakyusa (Tanzania)
Kinyarwanda (Rwanda)
Kitharaka (Kenya)
Kuria (Kenya)
Lamba (Zambia)
Luganda (Uganda)
Lusaamia (Kenya)
Makonde (Tanzania)
Ndebele (Zimbabwe)
Nyamwezi (Tanzania)
Sukuma (Tanzania)
Swahili
Swati (Swaziland)
Zulu (South Africa)

Fedor Rozhanskiy

Non-canonical behavior of reduplicated ideophones in Komi

Abstract: The paper considers reduplicated ideophones in the Komi language and is based on field materials collected by the author in the village of Eremeevo (Troitsko-Pechorsky District). Accent is placed on three aspects. First, Komi ideophones demonstrate a typologically rare combination of both progressive and regressive directions of reduplication. Second, they raise a question of the border between a root and an affix in sound-symbolic words, and the problem of morphological homonymy. Third, they form a continuum of constructions that stretches from typical morphologically bare ideophones to regular words of the ideophonic origin.

Keywords: reduplication, ideophones, Komi, morphophonology, syntax

1 Introduction

It is a well-known fact that ideophones often have reduplicated forms.[1] In a language with a developed ideophonic system, reduplicated ideophones usually constitute the major part.[2] Ideophonic reduplication often demonstrates interesting features that are not typical for canonical reduplication. For example, in Donno-So there are many triplicated ideophones with alternating vowels: *gɛngú-gangu-gɛngu* 'sparkling, flickering', *poró-para-porò* 'at a fast pace (about many people)' (Kervran 1982).

Ideophonic systems are found in many Uralic languages: Permyak (Krivoshchekova-Gantman 1964), Mari (Gordeev 1981, Rozhanskiy 2002), Mordvin (Imajkina 1968), Enets (Kazakevich 1994), Selkup (Kazakevich 1990), etc. Alt-

1 Cf., for example, Bartens (2000: 42): "... there is one morphonological process so characteristic of them [i.e., ideophones – F.R.] that it can often be used to identify ideophones: reduplication/reiteration".

2 E.g., Diakite (1989) gives the wordlist of Bamana ideophones that contains less than 250 non-reduplicated and more than 400 reduplicated entries.

Fedor Rozhanskiy: Institute of Estonian and General Linguistics, University of Tartu, Jakobi 2-410, 51014 Tartu, Estonia – Institute of Linguistics, Russian Academy of Sciences.
E-mail: handarey@yahoo.com

https://doi.org/10.1515/9783110599329-006

hough Uralic ideophonic systems are not as rich as in African languages, they demonstrate typologically rare and remarkable features.

This paper is a case study of ideophones in the Komi language. It addresses three untypical and/or theoretically challenging features of Komi ideophones: (1) the direction of reduplication, (2) the ambiguity in the morphological structure of ideophones, and (3) the untypical morphosyntactic properties of the ideophones.

To begin with, I will explicitly define several notions important for the current paper, as the relevant terminology can vary in different sources.

Total reduplication can be **exact** or **inexact**. In the first case, both parts of the reduplicated word are identical; in the latter, they are not.[3] For instance, the English *tick-tack* is inexact reduplication.

The original part of a reduplicated word is called a **base**; the other part is a **copy** of the base (see Section 3 that describes how to distinguish between the base and the copy). In the English *tick-tack*, the first part (*tick*) is the base and the second (*tack*) is the copy.

Reduplication is **progressive** if the copy follows the base; otherwise, it is **regressive** (the more popular terms right vs. left reduplication denote the same). For instance, *tick-tack* is an example of progressive reduplication.

The segmental difference between the copy and the base can be treated as alternation. A phonetic segment in the copy that corresponds to a different segment in the base is called a **replacing** vowel or consonant. For example, *a* is a replacing vowel in *tick-tack*. A scheme representing the alternation of segments will be called a **divergence pattern**. In *tick-tack*, the divergence pattern is *i-a*.

If two independent words that are rather close phonetically concatenate into one, the resulting complex can resemble a reduplicated form. In fact, it is not because this complex results from concatenation but not from copying. Such combinations are called **paired words**, cf. in Komi *tol'gi̮ni̮-bol'gi̮ni̮* 'to prattle' (lit. 'to chirp-talk'), *ge̮rni̮-ke̮ʒni̮* 'to work in the field' (lit. 'to plough-sow') or *pi̮rni̮-petni̮* 'to go-in-go out' (Kokkonen 2001: 117–118).

An **ideophone** is a representative of the class of ideophones. This class is defined on the basis of several criteria.[4] Not all ideophones necessarily conform

3 Identifying reduplication as exact or inexact can be more problematic in case of partial reduplication. For the current paper, this problem is not relevant.

4 In any language one can find a word (or several words) that resembles an ideophone. However, one word is not enough for claiming that there are ideophones as a morphological class in this language. For this reason, the listed criteria are applicable to the ideophones as a class but not to a separate word.

to all the criteria (features of ideophones vary cross-linguistically), but usually most of the criteria are fulfilled. These criteria are:[5]

1. Phonetic structure:

Ideophones have specific phonetic features. First, they often consist of a definite set of phonemes and their combinations.[6] Second, ideophones usually constitute several groups: each group has a specific structure and rules that determine the position of particular phonemes. Table 1 sums up the phonetic features of reduplicated ideophones with the $C_1V_1C_2C_3$-$C_1V_2C_2C_3$ structure in three languages that will be discussed below: Khakass, Mari and Komi.[7]

Table 1: Ideophones with $C_1V_1C_2C_3$-$C_1V_2C_2C_3$ structure in Khakass, Mari and Komi.

Language	Vowels		Consonants			Examples
	V_1	V_2	C_1	C_2	C_3	
Khakass	u, ɨ, î	a	any[8]	r	t	*sîrt-sart* 'in a magpie style', *purt-part* 'about bubbling porridge; about chatting', *xɨrt-xart* 'about cackling (of a hen)'
Mari	ə	o, u	any	r, l, š	t	*kərt-kort* 'fast, quick', *jəšt-jušt* 'quick', *šəlt-šolt* 'with an even knock (usually on metal objects)', *kəlt-kolt* 'ticking (about a pendulum clock and similar sounds)'
Komi	u	a, o, ȩ, e	any	r, l, l'	k, ć, s	*buls-bols* 'about slopping through the mud', *kurć-karć* 'about a crackle (of a cracking branch, etc.)', *purk-park* 'quick, in haste', *śurk-śark* 'with a cracking sound'

2. Specific meaning:

First, even in genetically distant languages, most ideophones belong to a definite set of semantic fields (noises, specific colors or shapes, irregular walks, intensive

5 I compiled my own set of criteria based on the data I have and on the experience of previous researchers. The definition and features of ideophones were discussed by many scholars; see, for example, Zhurkovskij (1968), Schaefer (1984), Dingemanse (2011: 21–55).

6 Cf. Stevenson (1969: 179) about ideophones in Bagirmi (one of the Central Sudanic languages): "Great use is made of reduplication, and it is noteworthy that their phonetic structure is unlike that of other words – a characteristic form is consonant + vowel + consonant, the final consonant being foreign to nominal, pronominal or verbal roots."

7 Khakass is a Turkic language; Mari and Komi are Finno-Ugric languages.

8 By "any" I mean that there are no obvious restrictions for this consonant.

or attenuative action, etc.). Second, it is common for ideophones to have a semantic structure untypical for other words. A regular word usually has rather abstract semantics (e.g., a table is a table, whatever the shape or size or color), and all the details concerning the denoted object come from the syntactic and pragmatic context. In contrast to that, the meaning of an ideophone can be expressed through a multimodal picture of a concrete situation. This picture can have various modes: visual, audible, olfactory ones, etc. and describes many details of the situation, cf. Krivoshchekova-Gantman (1964: 115); Samarin (1967).

3. Paronymic organization:
A significant part of ideophones constitutes groups of words with a similar but not identical phonetic shape, and with a meaning from the same semantic field. This effect comes from the sound-symbolic nature of ideophones: small differences in phonetic shape correlate with minor variations in the meaning.

4. Specific syntactic behavior:
Usually, an ideophone is either not integrated into the syntactic structure of the sentence or combines with a special auxiliary verb (see Section 5). The position in a sentence can also distinguish an ideophone from other words.[9] However, a language can have different subclasses of ideophones with their own syntactic behavior (Newman 1968).

5. Restricted usage:
Ideophones appear mostly in colloquial speech. In written texts, they are very rare, common only for specific genres, and serve as stylistic coloring of the text.[10]

6. Specific prosodic marking of ideophones:
This characteristic is well described for some groups of languages (see e.g., Zhurkovskij 1968: 30, Childs 2003: 118–119), but it was not studied accurately in the Uralic languages.

I would like to note that I do not use the sound-symbolic nature of ideophones as an independent distinctive feature. First, it is often difficult to prove that a particular word is sound-symbolic. Second, when an ideophone converts into another part of speech, its sound-symbolic origin remains; so, there are many words that have sound-symbolic origin but are not ideophones anymore.[11]

---

9 For instance, in the Mari language ideophones can formally be confused with adverbs. However, the most typical position for ideophones is before the auxiliary verb, while adverbs usually take other positions (often the initial position in the sentence).

10 Cf. for example Zhurkovskij (1972: 193–194) on Hausa ideophones: “Especially often Hausa ideophones appear in spontaneous speech and in fiction. <...> Business or official style excludes the use of ideophones almost completely <...>”.

11 See Hinton et al. (2006: 5) on the conventional sound symbolism.

Ideophones are present in many Uralic languages, but since most of these languages are situated on the territory of Russia where the Russian linguistic tradition has been dominating, in most grammars, ideophones are not treated as a separate part of speech. Very often, they are listed among adverbs and/or interjections, see, for example, Zhilina (1985: 92), Nekrasova (1998: 250) on Komi. Some scholars considered ideophones as a separate part of speech (e.g., Krivoshchekova-Gantman 1964), but this approach did not become dominating. Consequently, the study of Uralic ideophones is still in an initial stage.[12]

The structure of this paper is the following. The data and research methods are described in Section 2. Section 3 considers the direction of reduplication in Komi ideophones. The analysis is based on vowel alternations and derivation processes. Section 4 discusses the problem of morphological homonymy that is important for the morphosyntactic analysis of ideophonic constructions presented in Section 5. Conclusions are formulated in Section 6, and Section 7 gives the acknowledgements.

2 Data and methods

Data on the Komi language were collected in the village of Eremeevo (Troitsko-Pechorsky District in the Komi Republic) in 2002–2003; the field trips were organized by the Department of Theoretical and Applied Linguistics of Lomonosov Moscow State University. For the research, native speakers of different age groups were interviewed, but most data were collected from the older speakers. Among the teenagers, no one knew more than a few ideophones.[13]

Besides the Komi data, I refer to Mari and Komi material. The comparison of the three languages helps to demonstrate some specific features of Komi ideophones. I used the same collecting methods for all three languages to increase the comparability of the data.

The Meadow Mari data were collected in the village of Staryj Torjal (Novyj Torjal district in the Mari El Republic) in 2000 and 2001; the field trips were organized by the same department.

12 Obviously, the problem of ideophones as representatives of a particular word class is not specific for the Uralic languages. See, for example Dwyer & Moshi (2003).

13 The interviewed teenagers were fluent speakers, and Komi was their main language of communication in the village. This proves that the ideophonic system is one of the most fragile parts of the language that can be easily lost.

The Khakass data were collected in the village of Kazanovka (Askizsky District in the Republic of Khakassia) in 2001 and 2002; the field trips were organized by the Institute for Linguistics of Russian State University for Humanities.

As all the data used in this paper are dialectal, some differences from the corresponding standard language are possible.

It is widely known that collecting ideophones is not easy: they do not appear in elicitations, and they rarely appear in most text genres, with the exception of informal communication between native speakers.[14] For the current research, the following collection methods were used. A preliminary list of ideophones was compiled on the basis of the dictionaries (Lytkin 1961 for Komi, Baskakov & Inkizhekova-Grekul 1953 for Khakass, Serebrennikov 1956 for Mari). The native speakers were asked if they knew the words from the list. In case of a positive answer, a native speaker was asked to construct a sentence with this word. Next, I asked additional questions about the meaning, co-occurrence with the auxiliary verbs, and other semantic and grammatical features. After the native speaker realized what kind of words was the main point of interest, I asked for more examples of "this kind" of words.[15] As far as possible, the collected data were later discussed and checked with other native speakers.

3 Direction of the Komi reduplication

For exact reduplication, the direction cannot be defined (sometimes, it can be determined through indirect evidence). For inexact reduplication, there are at least two methods to define the direction.

The first method compares the two parts of the ideophones. Often, the vowels in the copy are different from those in the base. Consonant(s) can also change, but it happens much rarer in ideophones[16], and I do not discuss it in

14 "Even though one may be sure that the lexicon of a particular language includes a great many ideophones, they are extremely difficult to elicit, and they are often avoided by native speakers when speaking to someone whose competence in their language is clearly at a low level" (Welmers 1973: 461).

15 Often, the examples given by the native speakers contained paired words, but not the ideophones. Obviously, these examples were not included in the list of ideophones.

16 In other word classes, there are productive types of reduplication with consonant mutation, e.g. the well-known m-reduplication that is typical in Turkic and some other languages of the Middle East (Sofu 2005, Stolz 2008: 116–120, Chirikba 2008: 55–56).

this paper. By detecting the most frequent vowel(s), it is possible to define the replacing vowel and thus determine the direction of reduplication.

Based on the data from Tables 2 and 3, I will illustrate how this method works for Mari and Khakass languages.[17]

Table 2: Vowels in Mari reduplicated ideophones.

Vowel in the first part	Vowel in the second part	Quantity absolute and %	Examples
ə	o	28 (~67 %)	*jəlt-jolt* 'glimmering, sparkling', *rəšt-rošt* 'on the firm tread, striking by foot with force'
ə	u	8 (~19 %)	*jəst-jušt* 'quick', *kədər-kudər* 'crooked, clumsy'
ə	a	2 (~5 %)	*n'əmər-n'amər* 'viscous, sodden', *kədər-madər* 'crooked'
o	u	2 (~5 %)	*ola-vula* 'spotted', *joli-juli* 'glittering'
ə	ü	1 (~2 %)	*rəž-rüž* 'together, simultaneously'
a	u	1 (~2 %)	*kadər-guder* 'sinuous, winding'

Table 2 shows that 4 out of 6 divergence patterns of inexact reduplication in Mari have *ə* in the first part of the word. These patterns cover 93 % of all examples of reduplicated ideophones. It is easy to conclude that *ə* is a replacing vowel, and that Mari ideophones prefer the regressive reduplication (the second part of the word is the base, and the first part is the copy).

In Khakass (Table 3), 5 out of 6 divergence patterns have *a* in the second part and cover 79 % of all examples. Apparently, *a* is a replacing vowel, and the reduplication of Khakass ideophones is progressive (the first part of the word is the base and the second part is the copy).

However, the Komi ideophones (Table 4) are not similar either to Mari or to Khakass. On the one hand, 4 out of 7 divergence patterns have *u* in the first part of the ideophone and cover 80 % of examples. On the other hand, 3 out of 7 patterns have *a* in the second part of the ideophone; these patterns cover 66 % of examples.

17 The total number of attested ideophones with inexact reduplication is 42 for Mari, 62 for Khakass and 85 for Komi.

Table 3: Vowels in Khakass reduplicated ideophones.[18]

Vowel in the first part	Vowel in the second part	Quantity absolute and %	Examples
ɨ	a	33 (~53 %)	*sɨr-sar* 'about a loud and shrill crying of a child', *pɨlčix-palčix* 'floundering about in the water (about children, fish, etc.)'
a	u	13 (~21 %)	*taltaŋ-tultaŋ* 'about an unsteady pace of a baby', *sala-sula* 'anyhow, haphazardly'
u	a	9 (~15 %)	*xujbaŋ-xajbaŋ* 'about zigzag movements', *suum-saam* 'about a noise, hubbub'
o	a	3 (~5 %)	*soxɨr-saxɨr* 'many-colored, motley', *tox-tax* 'about a loud repetitive knock or rattle (wheels, hammer, etc.)'
î	a	2 (~3 %)	*sîîx-saax* 'about squeak, squeal or scream (e.g., of children or piglets)', *sîrt-sart* 'in a magpie style'
ɨ+e	a+a	2 (~3 %)	*sɨreŋ-saraŋ* 'about a frivolous woman', *sɨrex-sarax* 'about a jackanapes'

There is no big contrast between the "first *u*" and "second *a*" groups of reduplicated words; the size of the groups is comparable. Additionally, more than half of the examples have *u* in the first part and *a* in the second part of the ideophone, and therefore, correspond to both groups. We have to assume that there is a mixture of both progressive and regressive reduplication in Komi.

This hypothesis can now be tested with the second method of defining the direction of reduplication. This method relies on the assumption that the base of reduplication is more involved in the derivational processes than the copy. For the English word *tick-tack*, there is a verb *to tick*, but no verb *to tack* (in the corresponding meaning).

18 In Table 3, I do not draw a distinction between short and long vowels for naming the divergence patterns. In disyllabic words, the alternation usually concerns the vowel of the first syllable. The last divergence pattern involves both the first and the second vowel.

Table 4: Vowels in Komi reduplicated ideophones.

Vowel in the first part	Vowel in the second part	Quantity absolute and %	Examples
u	a	47 (~55 %)	*nuza-naza* 'whining, twanging', *śuľ-śaľ* 'with splashing'
u	o	17 (~ 20 %)	*škuv-skov* 'about a loud resonant sound (e.g., a gun shot)', *buľ-boľ* 'with gurgle, murmur'
i	o	8 (~9 %)	*viľski-voľski* '(to walk) sliding', *giľ-goľ* 'with a loud sonorous noise'
i	a	5 (~6 %)	*girska-garska* 'shuffling or scraping', *ǯiža-ǯaža* 'with hissing (e.g., while frying)'
i̮	a	4 (~5 %)	*gri̮ma-grama* 'rattling, rumbling', *ki̮ša-kaša* 'about rustling'
u	e	3 (~4 %)	*vuži̮n'-veži̮n'* 'wry, crooked', *ľutki-ľetki* '(to walk) staggering'
u	e̮	1 (~1 %)	*zurk-ze̮rk* 'about jerky movements'

We can conclude that *tick* is the base of reduplication and *tack* is the copy. Sometimes, both the copy and the base are involved in derivational processes in the same extent. However, I have not found any evidence that in some language the copy is systematically used for derivation but the base is not. Therefore, the second method is to find a pair of an ideophone and a cognate non-reduplicated word from another morphological class. The part of the ideophone, which is similar to the cognate word, is the base. If we know the base, we know if the reduplication is progressive or regressive. By examining many such pairs we can define which direction of reduplication is typical for ideophones in a particular language.

This method is applicable only if two conditions are met. First, a number of ideophones should have non-reduplicated words that are cognate with the base of the ideophone (occasional examples cannot serve as a reliable indicator of the direction of reduplication). Second, only the base should have cognate words. It is impossible to define the direction if both the base and the copy have cognates. There are two possible causes of the latter situation: (a) misinterpretation of the word as reduplicated instead of considering it a paired word, (b) intensive derivational processes that involve both parts of reduplicated words (however, this situation does look very typical cross-linguistically).

In Mari, most ideophones do not have cognate words, so this method cannot be applied. On the contrary, many ideophones in Khakass have cognates. With the exception of a few examples, the first part of an ideophone has a cognate but the second part does not, e.g., *tɨr-tar* 'about sound of crackling, chirring' (*tɨrlîrge* 'to cracle, to chirr', **tar-*), *pitɨr-patɨr* 'into smithereens, in different directions' (*pitɨrîrge* 'to scatter, to disperse', **patɨr-*), *nizɨrt-nazɨrt* 'about loud but not sonorous beating sounds' (*nizɨrîrge* 'to make a noise, to rattle', **nazɨr-*), *sibɨx-sabɨx* 'about wispering' (*sibɨxtîrge* 'to wisper', **sabɨx-*). The analysis of cognate words fully supports the conclusion that the Khakass ideophones demonstrate progressive reduplication.

Komi reduplicated ideophones require a more detailed analysis. Within the divergence patterns *u-o* and *u-e*, there are many ideophones whose second part has cognate words but no ideophones with cognates for their first part. For instance, *gul'-gol'* 'on sonorous sound (e.g., of a broken jar)' (*gol'* 'clank, tinkle', **gul'-*)[19], *gun'-gon'* 'looking with caution (turning one's head from side to side)' (*gon'* 'craning one's neck', **gun'-*), *vuži̮n'-veži̮n'* 'wry, crooked' (*veži̮n'-* '1) wry, crooked, 2) distortion', **vuži̮n'*), *l'ug-l'eg* 'wobbling, staggering' (*l'egni̮* 'to wobble, to stagger', **l'ug-*). This indicates that [u] is a replacing vowel, and the listed examples represent regressive reduplication.

However, the ideophones with the divergence patterns *i-a* and *i̮-a* have cognates that are similar to the first but not the second part, e.g., *gi̮ma-gama* 'on rumbling or loud knocking' (*gi̮mavni̮* 'to thunder', **gama-*), *gri̮ma-grama* 'rattling, rumbling' (*gri̮mgi̮ni̮* 'to rattle, rumble, roar', **grama-*), *giǯa-gaǯa* 'scratching, crunching' (*giǯgi̮ni̮* 'to scratch, to make crunching noise', **gaǯa-*).

As for the most widespread divergence pattern *u-a*, it has cognates with both parts. In most cases, the second part of the ideophones is involved: *lujk-lajk* 'waving, swinging' (*lajkn'itni̮* 'to rock, to swing', **lujk-*), *puś-paś* 'into smithereens' (*paśkavni̮* 'to spread', **puś-*), *nuza-naza* 'whining, twanging' (*nazgi̮ni̮* 'to whine, to twang', **nuza-*).[20] Still, there are examples of ideophones that only have congnates with their first part: *l'uź-l'aź* 'shaggy, tousled' (*l'uźgi̮ni̮* 'to stream, to trail', **l'aź-*), *mura-mara* 'about growling (e.g., of a bear)' (*muravni̮* 'to growl',

19 Here and below the indication that the word does not exist means that there is no such word in the corresponding meaning (the homonyms are not taken into account).

20 In Beznosikova et al. (2000), there is an entry *nuzgi̮ni̮* with the reference to *nazgi̮ni̮* 'to whine, to twang'. I consider it as an innovation (possibly dialectally restricted) derived from *nuza-naza*, because this verb is absent from Lytkin (1961) and it was not known to any of my language consultants.

**mara-*), *n'uv-n'av* '(noise of) something burning down fast' (*n'uvni̯* 'to lick', **n'av-*).[21]

On the basis of these observations, it seems that Komi ideophones should be divided into two main groups. One group is built through progressive reduplication and uses *a* as a replacing vowel. The other group is built through regressive reduplication and uses *u* as a replacing vowel.[22] For the ideophones with the divergence pattern *u-a*, it is impossible to define the direction of reduplication through the alternating vowels, but it can often be done through the analysis of cognate words.

4 Morphological structure of ideophones

Generally, morphological parsing (dividing a word into morphemes) is not a very difficult task. The main principle of defining a morpheme is transparent: if several forms have a similar sequence of phonemes and have something common in their meaning, this sequence of phonemes is probably a morpheme. For instance, the English forms *books, boots, tables, windows* end in *s* and have a plural meaning. Most probably, *s* is a plural suffix. The roots of non-cognate words are not similar. If it happens that roots in two non-cognate forms have some similarity (cf. *books* and *boots*), it should be considered a coincidence.

Typical difficulties of morphological parsing arise when drawing a border between morphemes in case of fusion, or when defining a formative as a separate morpheme or as a part of root in case of lexicalization. However, ideophones (and Komi ideophones, in particular) pose a number of more challenging problems.

Since ideophones are sound-symbolic words, the phonemes within ideophones often have a semantic interpretation. Usually, many ideophones consti-

21 In Beznosikova et al. (2000), there is a verb *n'avmunni̯* 'to catch fire'. This verb is absent from Lytkin (1961). Again, I consider it an innovation (possibly dialectally restricted).

22 Besides these two groups, there is a group of ideophones that have neither *u* in the first part nor *a* in the second (the divergence pattern *i-o*). This group is small and heterogeneous: there are ideophones that have cognates for the both parts or only for the second part: cf. *vil'ski-vol'ski* '(to walk) sliding' and *vil'sjavni̯* 'to glide, to slide', *vol'ski̯ni̯* 'to slip'; *tić-toć* 'tick-tock' and *tićki̯ni̯* 'to tick (colloquial)', *toćki̯ni̯* 'to tick'; *gil'-gol'* 'with a loud sonorous noise' and *gol'* '(onomatopoeic) clank', but not **gil'-*. However, there are no examples when only the first part has cognates. Therefore, this group of ideophones demonstrates regressive rather than progressive reduplication.

tute paronymic clusters, i.e., bundles of ideophones that have common features on both formal and semantic levels. Compare, for example, Komi ideophones *luč-lač* 'with creak, crackle, crunch', *ruč-rač* 'with loud crackle' and *tuč-tač* 'with clatter' or *bul'-bol'* 'with gurgle, murmur', *gul'-gol'* 'with clink, clang', *źil'-źol'* 'about sonorous sound (of a small bell or tiny brook)' and *śul'-śal'* 'with splashing'. It is obvious that at least the final consonant is not meaningless. This is principally different from the usual words where the root is semantically indivisible, and a phoneme (or a sequence of phonemes) that has a semantic interpretation is likely to be a morpheme.

Adding another phoneme to an existing ideophone can result in a new form with a slightly modified meaning. For example, the Komi ideophone *girska-garska* 'shuffling or scraping (about sound)' can be considered either a derivation from the ideophone *gir-gar* 'with a grinding or rumbling sound' (the suffixes *-s* and *-ka* of the "imitative adverbs" are mentioned in Saharova 1949: 36) or an underived word where *k* and *a* are sound-symbolic segments of the ideophonic root. From my point of view, there are no really strong arguments that can unambiguously define whether a form is independent or a derivative. Simplified solutions of the kind "if there is *k* it should be a suffix" do not work, cf. *lujk-lajk* 'waving, swinging' where the final *k* is a part of a root (first, there is no word **luj-laj*, second, it is not typical for Komi ideophonic roots to end in *j*).

Even if one can decide that some part of a Komi ideophone is definitely a suffix, one cannot be sure which suffix it is, because next comes the problem of homonymy. Homonymy concerns several suffixes. For example, the suffix *-a* is (a) one of the suffixes of imitative adverbs (ideophones in my terminology) (Saharova 1949: 36), (b) an adverbial suffix that derives adverbs from different stems (Lytkin 1955: 249), (c) a verbal suffix that marks 1SG of the present and future tenses (Lytkin 1955: 212, 218).

Let us consider examples (1) and (2) where the same ideophonic root is combined with different suffixes. There is no doubt that in example (1) the ideophone *bulskis-bolskis* contains the verbal marker *-is* (the 3SG of the past tense).

(1)	*jęra*	*n'ur-as*	*bulskis-bolskis*
	elk(NOM)	marsh-INESS.POSS.3SG	IDEO(PST1.3SG)
	'An elk was squelching in the marsh'		

However, it is more difficult to give a morphological interpretation for the form *bulska-bolska* in example (2): does it contain an adverbial suffix, an ideophonic suffix, or is *a* a part of the root?

(2)	*je̮ra*	*n’ur-as*	*bulska-bolska*	*ki̮l-i-s*
	elk(NOM)	marsh-INESS.POSS.3SG	IDEO	hear-PST1-3SG
	‘An elk was squelching in the marsh’			

In example (3), *a* is combined with a different ideophonic root. Formally, it can be interpreted as an adverbial suffix, an ideophonic suffix, a part of the root, or a verbal suffix of the 1SG present tense.

(3)	*me*	*viľska-voľska*	*mun-a*	*tuj*	*kuźa*
	1SG(NOM)	IDEO	go-PRS.1SG	road(NOM)	along
	‘I am walking along the road sliding’				

Examples (4) and (5) illustrate the homonymy of *i*. In both sentences, the ideophone *viľski-voľski* follows the verb, so the construction does not look like the “ideophone+auxiliary” combination. Formally, *i* could be a verbal marker of the 3SG past tense, but in this case, it is not clear why it appears in (4), where the action is in the present tense. Still, in (5) it is possible to interpret the ideophone as a verbal form that specifies the meaning of the previous verb, i.e., ‘I was walking, sliding’. Apart from the verbal interpretations, both in (4) and in (5) *i* can be interpreted as a part of the root, an ideophonic suffix (in a strange construction where the ideophone follows the verb), or an adverbial suffix which was not mentioned in grammars.

(4)	*me*	*mun-a*	*viľski-voľski*
	1SG(NOM)	go-PRS.1SG	IDEO
	‘I am walking sliding’		

(5)	*me*	*mun-i*	*viľski-voľski*
	1SG(NOM)	go-PST1.1SG	IDEO
	‘I was walking sliding’		

From my point of view, the whole system of suffixes proposed in (Saharova 1949) should be reconsidered. Saharova (1949: 36) lists the following suffixes of imitative adverbs (ideophones in my terminology): *-a, -ka, -i, -ki, -s*. She also tries to describe their semantic features (*-ka* and *-ki* express “multiple action”; *-s* expresses “quickness of sounding of an action or swiftness (instantaneity) of a movement”; *-a* and *-i* “convey diversity of nuances of a sound produced by an action”).[23] In my opinion, there are four but not five suffixes: *-a, -k, -i* and *-s* (*-ka* and *-ki* are combinations of *-k* with *-a* or *-i*).

23 I am not going to discuss the meaning of these suffixes, but I would like to note that a number of my field examples did not demonstrate the semantics described by Saharova (1949).

Saharova (1949: 40–41) further lists suffixes that derive verbs from imitative words. The most frequent are *-gi̮*, *-ki̮*, *- i̮*, *-si̮* (*-gi̮* and *-zi̮* appear after a voiced consonant; *-ki̮* and *-si̮* after a voiceless consonant). The vowel *i̮* in these suffixes is an epenthetic sound that is inserted to avoid long consonant clusters; therefore, these suffixes will be spelled as *-k*, *-g*, *-z*, *-s* below (one can find such spelling also in Lytkin 1955: 249).

Following this analysis, we have to accept homonymic variants of *k* and *s*. On the one hand, *k* and *s* are suffixes of imitative adverbs (Saharova 1949: 36), on the other hand, they are verbalizers that derive verbs from ideophones (Saharova 1949: 40–41, Lytkin 1955: 249).

In my opinion, there is no need to distinguish ideophonic suffixes as such. They only make the description more complicated. Instead, I suggest introducing segments that have a sound-symbolic interpretation in ideophones, in addition to the classical derivational and inflexional suffixes. These segments should not be part of the morphological system, but rather a specific feature of ideophones. However, in each particular case it is not always easy (or even possible) to define, which of the three entities (ideophonic segment, inflectional suffix or derivational suffix) we are dealing with.

5 Morphosyntactic properties of Komi ideophones

In this section, I would like to demonstrate that the margins between Komi ideophones and other parts of speech are blurred.

In most languages, ideophones are morphologically bare words, i.e., they do not have any morphological markers.[24] Very often, an auxiliary verb is used to integrate an ideophone into the syntactic structure of a sentence. Without an auxiliary, ideophones do not entirely fit into normal syntactic patterns (Voeltz & Kilian-Hatz 2001: 2).

There are three most typical strategies of the syntactic behavior of ideophones:

---

24 "A salient feature that distinguishes ideophones from many other words is that there is hardly any affixation in the morphology of this group of words" (Bodomo 2006: 204). However, cf. Zhurkovskij (1968: 16): "The feature of null-morphology, which seems to be useful for ideophones in most languages, should be adjusted concerning ideophones in some other languages, mostly agglutinative".

1. An ideophone is a morphologically bare word with a low integration into the syntactic structure[25] (often, it looks like something between an interjection and an adverb).
2. A morphologically bare ideophone combines with an auxiliary verb that bears morphological markers. Typical auxiliaries are "to do", "to say", "to become", "to hear", etc. The resulting complex "ideophone + auxiliary verb" serves as the predicate of a sentence, see, for example, (Shay 2014: 581).
3. An ideophone loses reduplication, attaches inflectional markers and in fact turns into some other part of speech (usually, a verb).

All three strategies are found in Komi.

Strategy 1. Bare ideophonic root.

(6) *ǯoǯ-ti* *ki̮skal-e̮ni̮* *ǯek* *gir-gar*
floor-PROL drag-PRS.3PL chair(NOM) IDEO
'A chair is dragged with noise on the floor'

(7) *ćeri* *ćuke̮r-i̮s* *kaćajtć-e̮ni̮* *buv-bov*
fish(NOM) bunch-POSS.3SG swing-PRS.3PL IDEO
'A bunch of fish is swinging back and forth'

These are examples of a typical ideophonic construction. A regular sentence that is syntactically and semantically complete adds an ideophone. The ideophone modifies the meaning of the sentence: it adds expressiveness and makes the meaning more concrete. The ideophone is poorly integrated into the syntactic structure of the sentence and is usually placed at the end.

Strategy 2. Combination of a reduplicated ideophone with an auxiliary verb. The most typical auxiliary verbs in Komi are: *munni̮* 'to go', *ki̮vni̮* 'to hear', *viǯ́ni̮* 'to hold', *kerni̮* 'to do', *vartni̮* 'to hit', *ve̮ćni̮* 'to do'.

(8) *oš* *ruč-rač* *mun-i-s*
bear(NOM) IDEO go-PST1-3SG
'A bear passed by with crackling'

(9) *śe̮le̮m-e̮j* *tip-tip* *ker-i-s*
heart-POSS.1SG(NOM) IDEO do-PST1-3SG
'My heart started beating'

25 "Across languages, ideophones tend to show a great measure of syntactic independence: they tend to occur at clause edges rather than deeply embedded within them; they tend to be aversive to inflectional morphology" (Dingemanse 2012: 656).

Within this strategy, an ideophone is a part of predication. An ideophone does not have morphological markers; it expresses only the lexical meaning. The grammatical properties are encoded in the auxiliary verb whose lexical meaning is diminished. The auxiliary verb follows the ideophone, and their syntactic tie is so strong that, in the tradition of Komi studies, such combinations were often considered as one word and spelled correspondingly (a verb with a "secondary suffix", cf. Lytkin 1955: 249). Ideophones can also combine with verbs that preserve their lexical meaning. In this case, an ideophone is spelled as a separate word, but it is not always clear what is the difference between this construction and the construction spelled as one word, cf. Fejes (2004: 6–7).

Strategy 3. The third strategy is not an ideophonic construction in the strict sense. It represents the way for an ideophone to become a common word. Most typically, an ideophone (a bare root) obtains a verbalizing suffix and becomes a regular verb. There is a special set of suffixes that derive verbs from ideophones in Komi (Saharova 1949: 40–41). In the verbalization process, an ideophone loses reduplication, because in a morphologically rich language, "normal" words are usually not reduplicated. For example, the verb *gri̯mgi̯ni̯* 'to rattle, rumble, roar' is derived from the ideophone *gri̯ma-grama* 'rattling, rumbling', the verb *l'aski̯ni̯* 'to throw with one's might; to stick on' is derived from *l'us-l'as* 'tumbling, plopping down', see more examples in Lytkin (1955: 249).

Apart from the specific derivational suffixes, the verbs of ideophonic origin do not differ from other verbs on the synchronic level[26], see (10).

(10)	*puška-jas*	*poms'a*	*gri̯mg-i-sni̯*
	cannon-PL(NOM)	constantly	roar-PST1-3PL
	'The cannons were constantly roaring'		[Lytkin 1961: 174]

As mentioned above, these three strategies are very typical cross-linguistically. However, in Komi we find at least four more strategies.

Strategy 4. A reduplicated ideophonic root adds an adverbial marker. In example (11), the ideophone *gura-gara* is reduplicated and has the adverbial suffix *-a* (the bare root form of this ideophone is *gur-gar*).

(11)	*ćeľad'*	*gura-gara*	*ve̯ć-i-sni̯*
	children(NOM)	IDEO	do-PST1-3PL
	'Children fumbled around'		

26 However, according to Ludykova (1992), such verbs can often be used in combination with a non-ideophonic verb in the same grammatical form, see the discussion below.

Strategy 5. A reduplicated ideophone gets adverbial markers and combines with a meaningful verb.

(12)

ni̯v	*mun-ę*	*jurśi-sę*	*ľuźa-ľaźa*	*leʒ́-ęm-a*
girl(NOM)	go-PRS.3SG	hair-ACC.POSS.3SG	IDEO	loose-PST2-3SG

'A girl is walking with tumbled hair'

In example (12), the ideophone *ľuźa-ľaźa* is reduplicated and has the adverbial suffix *-a* (the bare root is *ľuź-ľaź*), same as in the previous strategy. However, since *leʒ́ęma* is a meaningful verb but not a delexicalized auxiliary, this construction goes even further away from a typical ideophonic construction.

Strategy 6. A reduplicated ideophonic root adds verbal markers, but is still combined with an auxiliary verb.

(13)

pęćę	*lunti̯r*	*brutkis-brotkis*	*ki̯l-ę*
grandmother(NOM)	whole.day	IDEO(PST1.3SG)[27]	hear-PRS.3SG
ćeľad'	*vi̯lę*		
children(NOM)	on		

'Grandmother grumbled at the children the whole day'

This strategy seems to be the most complicated, because an ideophone is marked with verbal suffixes, but additionally an auxiliary verb is present.

It is interesting to compare this strategy with constructions described by Ludykova (1992). She describes combinations of ideophonic verbs with non-ideophonic verbs. An ideophonic verb appears after an ideophone loses reduplication and becomes a regular verb (cf. with Strategy 3). Non-ideophonic verbs that combine with ideophonic verbs can be both semantically bare auxiliaries and regular lexical verbs. According to Ludykova, the morphological characteristics of both the ideophonic and non-ideophonic verbs should be the same, e.g., *šporędćęni̯-lebalęni̯* '(They) flutter-fly', *ʒ́olźęni̯-śi̯lęni̯* '(They) chirp-sing' (Ludykova 1992: 96–97).[28]

However, in my examples with reduplicated ideophones, the morphological forms of the ideophone and the auxiliary are different. Moreover, some native speakers consider the examples where the forms of the two verbs are the same (14) as ungrammatical or questionable:

---

27 As discussed in Section 4, the morphological parsing of ideophones is often ambiguous. For this reason, morphological characteristics of ideophones are indicated in parenthesis without parsing.

28 From my point of view, such constructions are rather a type of paired words.

(14)

?/**pe̮će̮*	*lunti̮r*	*brutkis-brotkis*	*ki̮l-i-s*
grandmother(NOM)	whole.day	IDEO(PST1.3SG)	hear-PST1.3SG
ćeľad'	*vi̮le̮*		
children(NOM)	on		

'Grandmother grumbled at the children the whole day'

Strategy 7. A reduplicated ideophonic root adds verbal markers but loses an auxiliary.

(15)

oš	*ve̮r-i̮n*	*ručkis-račkis*	*kos*	*ul-jas-e̮n*
bear(NOM)	forest-INESS	IDEO(PST1.3SG)	dry(NOM)	branch-PL-INSTR

'A bear was making a noise by crackling dry branches in the forest'

(16)

me	*talun*	*brutki-brotki*	*ni̮l-ej*	*vi̮le̮*
1SG(NOM)	today	IDEO(PST1.1SG)	daughter-POSS.1SG(NOM)	on

'I grumbled at my daughter today'

(17)

pe̮će̮	*talun*	*vetle̮dl-e̮*	*da*
grandmother(NOM)	today	walk-PRS.3SG	and
lunti̮r	*brutke̮-brotke̮*		
whole.day	IDEO(PRS.3SG)		

'The grandmother is walking and grumbling the whole day'

In this case, both parts of a reduplicated ideophone have regular verbal markers that express tense, person, and number. Hence, each part of the ideophone looks like a regular verb, and the whole predicate reminds a repeated verbal form. Except for the reduplication, this strategy is similar to Strategy 3.

A dichotomous classification dividing all ideophones into two groups – primary and grammaticalized (see Dwyer & Moshi 2003) – is not enough to describe the behavior of the Komi ideophones. The seven possible Komi strategies are plotted on Figure 1. The figure shows how different constructions form a continuum that fills the space between a typical ideophonic construction and a regular word of ideophonic origin.

Based on Figure 1, I would also expect a class of regular Komi adverbs that are derived from ideophonic roots. However, it is much more difficult to distinguish between adverb-like ideophones and regular adverbs than between ideophones and verbs. In my data, I do not have clear examples of such adverbs.

In fact, the borders between the types listed in Figure 1 are also vague. In example (18), an ideophone that lost reduplication is followed by an auxiliary (this ideophone normally has a reduplicated form, and the reduplicated form would be natural in this context).

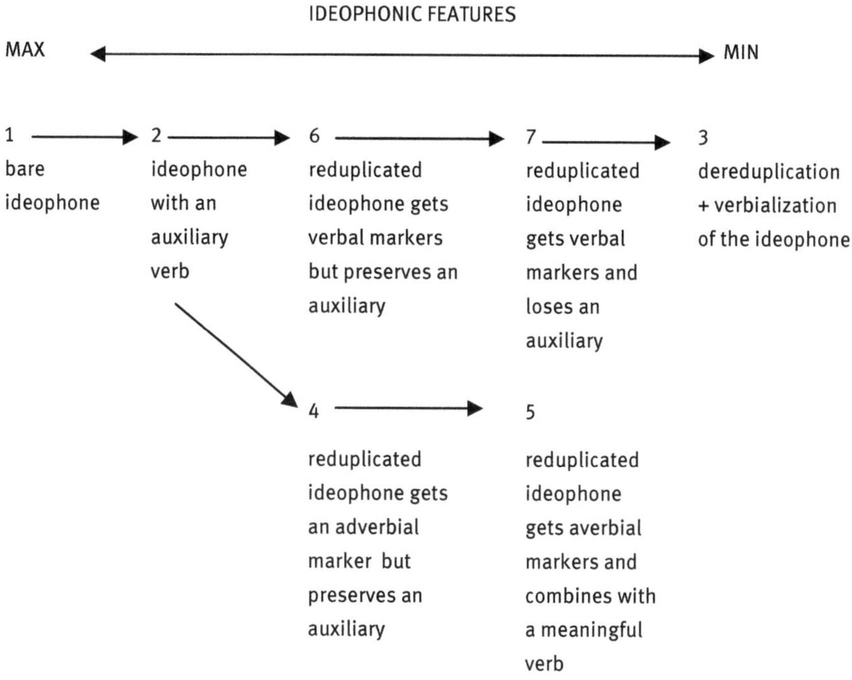

Figure 1: Continuum of Komi constructions with ideophonic roots.

(18)	*banka-i̯n*	*va-i̯s*	*bul'a*	*ki̯l-e̮*
	jar-INESS	water-POSS.3SG(NOM)	IDEO	hear-PRS.3SG

‘Water is splashing in a jar’

There are less ideophonic features in example (18) than in example (11), because there is no reduplication in (18). However, it is not clear if losing reduplication moves this construction closer to Strategy 2 (regular predication) or closer to Strategy 4 (an adverbial construction), compare with example 11 where the lack of an auxiliary makes the construction adverbial-like.

To a great extent, the problems with defining borders between different types of constructions appear due to the morphological homonymy discussed in Section 4. We have to deal with many ambiguous constructions that can be interpreted in different ways. I assume that this ambiguity is a feature of the Komi ideophones rather than the problem of choosing a correct linguistic interpretation.

6 Conclusions

This research inspected some features of reduplicated ideophones in Komi that are not typical cross-linguistically.

First, Komi has bidirectional inexact reduplication. Some Komi ideophones are built through regressive reduplication, and some through progressive. This feature of Komi ideophones is not typical from the typological point of view: usually, reduplication used in a particular function is omnidirectional (either progressive or regressive).

Second, Komi ideophones demonstrate volatile morphosyntactic behavior – from the most typologically common strategies to various transitional stages on the way to other parts of speech. Consequently, the borders between morphological classes become blurred. There are many constructions in which an ideophone acquires both morphological and syntactic features of a verb or adverb but does not fully convert into a different part of speech. This kind of behavior can be labeled as "morphological mimicry".

Third, the morphological structure of many Komi ideophones is ambiguous. Even though Komi is an agglutinative language and usually avoids homonymous markers, the morphology of ideophones often allows several interpretations. This ambiguity is rooted in the sound symbolic nature of ideophones. It is impossible to strictly distinguish between a semantically interpretable part of the root and a morphological marker of the same phonetic shape.

One possible explanation of these specific features is that a language with rich morphology has more problems with integrating reduplicated forms into the grammatical system than a morphologically bare language. The interpretation of reduplicated forms is always problematic from the morphological point of view, and they create an area of ambiguity where the degree of variation significantly increases.

Acknowledgments: This research was supported by the Estonian Research Council grant UIT2-37. I am deeply thankful to Nikolay Kuznetsov for checking the Komi examples.

Abbreviations

1, 2, 3	first, second, third person
ACC	accusative
IDEO	ideophone
INESS	inessive
INSTR	instrumental
NOM	nominative
PL	plural
POSS	possessive marker
PROL	prolative
PRS	present tense
PST1	the first past tense (imperfect)
PST2	the second past tense (perfect)
SG	singular

References

Bartens, Angela. 2000. *Ideophones and sound symbolism in Atlantic Creoles*. Saarijärvi: Gummerus Printing.

Baskakov, Nikolaj A. & Anastasija I. Inkizhekova-Grekul. 1953. *Hakassko-russkij slovarj*. Moskva: Gosudarstvennoje izdateljstvo inostrannyh i nacionaljnyh slovarej.

Beznosikova Ljucija M., Evgenija A. Ajbabina & Raisa I. Kosnyreva. 2000. *Komi-russkij slovarj*. Syktyvkar: Komi knizhnoje izdateljstvo.

Bodomo, Adams. 2006. The structure of ideophones in African and Asian languages: The case of Dagaare and Cantonese. In John Mugane, John P. Hutchison & Dee A. Worman (eds.), *Selected Proceedings of the 35th Annual Conference on African Linguistics*, 203–213. Somerville: Cascadilla Proceedings Project.

Childs, G. Tucker. 2003. *An introduction to African languages*. Amsterdam/Philadelphia: John Benjamins.

Chirikba, Viacheslav A. 2008. The problem of the Caucasian Sprachbund. In Pieter Muyksen (ed.), *From linguistic areas to areal linguistics*, 25–92. Amsterdam/Philadelphia: John Benjamins.

Diakite, Dzhibril. 1989. Ideofonicheskaja leksika v sisteme chastej rechi (na materiale jazyka bamana). Ph.D. thesis Institut Jazykoznanija AN SSSR, Moskva.

Dingemanse, Mark. 2011. The meaning and use of ideophones in Siwu. Ph.D. thesis University of Nijmegen.

Dingemanse, Mark. 2012. Advances in the cross-linguistic study of ideophones. *Language and Linguistics Compass* 6(10). 654–672.

Dwyer, David & Lioba Moshi. 2003. Primary and grammaticalized ideophones. In John Mugane (ed), *Linguistic typology and representation of African languages*, 173–187. Trenton/ Asmara: Africa World Press.

Fejes, László. 2004. "Compound verbs" in Komi: Grammaticalization without a grammatical morpheme? *Acta Linguistica Hungarica* 51. 5–43.

Gordeev, Fedor I. 1981. Podrazhateljnyje slova i etimologija. In *Voprosy marijskoj dialektologii*, 138–147. Joshkar-Ola: Marijskij nauchno-isledovateljskij institut.

Hinton, Leanne, Johanna Nichols & John J. Ohalam. 2006. Introduction: Sound-symbolic processes. In Leanne Hinton, Johanna Nichols & John J. Ohala (eds.), *Sound symbolism*, 1–12. Cambridge: Cambridge University Press.

Imajkina Marija D. 1968. Narechijno-izobraziteljnyje slova v mordovskih jazykah. Ph.D. thesis Mordovskij gosudarstvennyj universitet, Saransk.

Kazakevich, Olga A. 1990. Issledovanije izobraziteljnoj leksiki seljkupskogo jazyka. In Olga A. Kazakevich, *Ispoljzovanije EVM dlja issledovanija bespisjmennyh i mladopisjmennyh jazykov (Na materiale seljkupskogo jazyka). Uchebno-metodicheskoje posobije*, 30–68. Moskva: Izdateljstvo Moskovskogo universiteta.

Kazakevich, Olga A. 1994. Izobraziteljnaja leksika eneckogo jazyka. In *Znak. Sbornik statej po lingvistike, semiotike i poetike*, 203–212. Moskva: Russkij uchebnyj centr.

Kervran, Marcel. 1982. *Dictionnaire Dogon Donno Sɔ. Région de Bandiagara*. Bandiagara.

Kokkonen, Paula. 2001. A comparison of Komi serial verbs with those in Udmurt and Volgaic. In Tõnu Seilenthal (ed), *Congressus nonus internationalis Fenno-Ugristarum, 7.–13.8.2000 Tartu. Pars V. Dissertationes sectionum" Linguistica II*, 113–120. Tartu.

Krivoshchekova-Gantman, Antonina S. 1964. Mesto izobraziteljnyh slov v sisteme chastej rechi komi-permjackogo jazyka. In *Voprosy finno-ugorskogo jazykoznanja. Grammatika i leksikologija*. 112–121. Moskva/Leningrad: Nauka.

Ludykova, Valentina M. 1992. Izobraziteljnyje konstrukcii v komi jazyke. In. G. V. Fedjuneva (ed.), *Permistika: Dialekty i istorija permskih jazykov*, 92–100. Syktyvkar: Komi nauchnyj centr UrO Rossijskoj akademii nauk, Institut jazyka, literatuty i istorii.

Lytkin, Vasilij I. (ed.). 1955. *Sovremennyj komi jazyk. Uchebnik dlja vyshih uchebnyh zavedenij. Chastj pervaja. Fonetika, leksika, morfologija*. Syktyvkar: Komi knizhnoje izdateljstvo.

Lytkin, Vasilij I. (ed.). 1961. *Komi-russkij slovarj*. Moskva: Gosudarstvennoje izdateljstvo inostrannyh i nacionaljnyh slovarej.

Nekrasova, Galina A. 1998. Mezhdometije. In G.V. Fedjuneva (ed.), *Komi jazyk: enciklopedija*. 250–251. Syktyvkar: DIK.

Newman, Paul. 1968. Ideophones from a syntactic point of view. *Journal of West African languages* 4(2). 107–117.

Rozhanskiy, Fedor I. 2002. Reduplikacija v marijskom jazyke. In *Lingvisticheskij bespredel. Sbornik statej k 70-letiju A.I. Kuznecovoj*, 75–95. Moskva: Isdateljstvo Moskovskogo Universiteta.

Saharova, Marfa A. 1949. Izobraziteljnyje slova v komi jazyke. In D. V. Bubrih & P. N. Perevoshchikov (eds.), *Sovetskoje finnougrovedenije. Tom chetvertyj*, 33–42. Izhevsk: Udmurtgosizdat.

Samarin, William J. 1967. Determining the meaning of ideophones. *Journal of West African languages* 4(2). 35–41.

Schaefer, Ronald R. 1984. Towards an understanding of some ideophones of color in Emai. *Journal of West African languages* 4(2). 125–134.

Serebrennikov, Boris A. (ed.). 1956. *Marijsko-russkij slovarj*. Moskva: Gosudarstvennoje izdateljstvo inostrannyh i nacionaljnyh slovarej.

Shay, Erin. 2014. Afroasiatic. In Rochelle Lieber & Pavol Štekauer (eds.), *The Oxford handbook of derivational morphology*, 573–590. Oxford: Oxford University Press.

Sofu, Hatice. 2005. Acquisition of reduplication in Turkish. In Bernhard Hurch (ed.), *Studies on reduplication*, 493–509. Berlin: Mouton de Gruyter.
Stevenson, Roland C. 1969. *Bagirmi grammar*. Khartoum: University of Khartoum.
Stolz, Thomas. 2008. Total reduplication vs. echo-word formation in language contact situations. In Peter Siemund & Noemi Kintana (eds.), *Language contact and contact languages*, 117–132. Amsterdam/Philadelphia: John Benjamins.
Voeltz, F. K. Erhard & Christa Kilian-Hatz. 2001. Introduction. In F. K. Erhard Voeltz & Christa Kilian-Hatz (eds.), *Ideophones*, 1–8. Amsterdam/Philadelphia: John Benjamins.
Welmers, William E. 1973. *African language structures*. Berkeley/Los Angeles/London: University of California Press.
Zhilina, Tatiana I. 1985. *Luzsko-letskij dialekt komi jazyka*. Moskva: Nauka.
Zhurkovskij, Boris V. 1968. *Ideofony: sopostaviteljnyj analiz (na materiale nekotoryh jazykov Afriki i Evrazii)*. Moskva: Institut Jazykoznanija AN SSSR.
Zhurkovskij, Boris V. 1972. Tipologija ideofonov v hausa. In N. V. Ohotina (ed.), *Fonologija i morphonologija afrikanskih jazykov*, 157–195. Moskva: Nauka.

Thomas Stolz

(Non-)Canonical reduplication

Abstract: This study introduces canonical reduplication as a common yardstick for reduplicative phenomena as they are discussed for individual languages or cross-linguistically. The (mainly methodological) advantages of the canonical approach over extant competitors (as e.g. prototypical reduplication) in research dedicated to reduplication in general are outlined by way of ticking off a selection of the plethora of putative as well as bona fide instances of reduplication which figure prominently in the relative literature. The unclear phenomenology of reduplication can be shown to be ordered according to the general principles of canonicity. The canonical type of reduplication allows us to reflect about the legitimacy of some of the basic distinctions which have been made in reduplication-related studies traditionally.

Keywords: canonicity, prototypicality, continuum, non-canonical reduplication

1 Introduction

This study takes as its point of departure the confusingly wide range of phenomena which bear the label reduplication in the linguistic literature dealing either with the grammatical structures of an individual language or with the cross-linguistic distribution of reduplicative constructions.[1]

Since proponents of different theories and models disagree as to which of the cases are acceptable instances of reduplication (Stolz et al. 2015: 822–824), I employ the label potential reduplication (throughout Section 2.2) to refer indiscriminately to these cases without prejudicing in any way against or in favor of the tenability of their classification as instances of reduplication. Since the ex-

1 For reasons of space, we skip discussing the equally disturbing fact that there are also many alternative terms which are fully or partly synonymous with reduplication such as, e.g., (re-)iteration (Aboh et al. 2012: 1) and doubling (Inkelas 2005: 65). In some of these cases, reduplication is only a sub-category of a by far larger notion. It is certainly worthwhile scrutinizing the relation of reduplication to its all-embracing competitors. However, this is a big issue which deserves to be addressed in a book-length study separately.

Thomas Stolz: FB 10: Language Sciences, University of Bremen, Bibliothekstr., 28359 Bremen, Germany. E-mail: stolz@uni-bremen.de

https://doi.org/10.1515/9783110599329-007

act status of many of the examples of potential reduplication remains doubtful nevertheless, the term configuration is used in lieu of assigning the label construction to cases of potential reduplication automatically.[2]

To tidy things up in this domain, a prototype approach has been suggested by Stolz et al. (2015: 825–827). Section 3.1 is dedicated to the discussion of the notion of prototypical reduplication which presupposes that there are good and bad representatives of reduplication. The best exemplars are instances of prototypical reduplication which contrast with instances of non-prototypical reduplication. It is argued that in the case of prototypical reduplication it is justified to re-introduce the term construction. Non-prototypical reduplication, however, is often better described with reference to the notion of configuration. The prototype approach has its limitations. This is shown in Sections 3.2.–3.3 where the concept of canonical reduplication is focused upon. In these sections, the advantages of the canonical approach are argued for predominantly from a theory-based angle. Section 4 is devoted to a much more empirically oriented illustration of the relation between canonical reduplication and non-canonical reduplication with Section 4.1 challenging the interpretation that is given to the notion of total reduplication in the literature dedicated to this issue. Section 4.2 has the task of demonstrating how different instances of non-canonical reduplication can be ranked hierarchically on a scale of decreasing canonicity. Because of their relevance to the general topic of this edited volume, echo-word formations are given special prominence in this section. The conclusions in Section 5 sketch the lessons I have learned from applying the canonical approach to reduplication.

In terms of theory, I follow the lead of Stolz et al. (2011) as far as reduplication as such is concerned.[3] This means that, without adhering to *Optimality Theory*, I am working on the basis of axioms which are reminiscent of Steriade's (1988) *Full-Copy Model* (FCM) and the *Base-Reduplication Correspondence Theory* (BRCT) as outlined in Kager (1999). The particulars of this purposefully eclec-

---

2 I conceive of configurations as the co-occurrence of linguistic units (= the members of the configurations) in the same space which is defined by the limits of an utterance. Whether or not configurations may claim the status of collocations (Stefanowitsch & Gries 2005) is an interesting issue which cannot be raised in this study because no statistical analysis of corpora forms part thereof. Constructions on the other hand have constituents, an internal structure, a distinctive meaning/function, and are part of sentence-grammar.

3 In the very same study of the (supposed) universality of total reduplication, Stolz et al. (2011: 71–99) provide a survey of the history of thought in the domain of the linguistics of reduplication to which I have nothing in particular to add and thus refer the historically interested reader to this earlier publication.

tic approach of mine are exposed in Section 2.1 which gives an account of the basic notions and the terminology to which I adhere in this paper. Furthermore, I look at (prototypical and) canonical reduplication from the point of view of *Construction Grammar* (Fischer & Stefanowitsch 2006). For the notions of canonicity and non-canonicity, I am indebted to Corbett (2005). My exclusively qualitative methodology is that of functional typology (Croft 1993: 18–20) although this study is not a properly cross-linguistic survey. The data I present do not stem from a sample the members of which have been put together according to formal principles of optimal sampling as put forward, e.g., in Miestamo et al. (2016). The exclusive purpose of the sample is that of providing data which illustrate a given constellation of facts. The languages of this sample are listed in the Appendix. No attempt is made to quantify my findings at this point. In contrast to Rubino (2005a–b) and Mattes (2014: 32–33), I do not check the productivity of a given reduplicative phenomenon so that there are no estimates as to the type and/or token frequency of the cases under scrutiny. The data are analyzed strictly synchronically. Diachronic issues are relegated to future follow-up studies.

2 Basic concepts and facts

This section is divided in two parts which are meant to provide a suitable basis for the subsequent discussion of (non-)canonicity in Section 3–4. Section 2.1 therefore explains the categories I need to conduct the analytical procedure properly. This means that I do not only give content to the labels and notions I refer to but also expose my elementary convictions as to the nature of reduplication in general. Section 2.2 on the other hand is intended to give the reader an idea of the state of confusion that is created by the relatively common practice of lumping together many very diverse phenomena so that they indiscriminately count as representatives of reduplication. This situation calls for being remedied. I suggest a remedy in Section 3.

2.1 From my terminological and notional tool-box

In line with several approaches (as e.g. Goodwin Gómez & Van der Voort 2014: 2), I axiomatically assume that for reduplication to be the case there must be a base-copy relationship of two phonologically realized chains of segments, i.e. under reduplication, there is always a linguistic unit A in the function of base which is copied from and there is always another linguistic unit A' which constitutes the copy of this base. Thus, reduplicating A means copying A. Moreover, a particularly good copy A' ideally shares all properties with the base A. If A' resembles A only to a limited extent, however, the recognizability of A' as a copy of A may be impaired severely. Moreover, it is assumed that the copy bears resemblance to the base as much as possible for both the expression side and the content side. The base is copied purposefully so that the base-copy pair constitutes a distinct functional unit. Their forming part of one and the same functional unit favors the topological vicinity of base and copy. What these stipulations mean for canonicity is shown in Section 3.3.

In this context, Mel'čuk (1996: 42) introduces a set of notions which reveal themselves to be very helpful when it comes to describing reduplicative phenomena.[4] I present these notions according to the interpretation given to them by Stolz et al. (2011: 40) – with only minor additions applied for clarification.

- domain = the portion of a meaning-bearing unit within which reduplication applies;
- reduplicand (aka base) = the portion of the domain which serves as input for the copying process (the variable $A_{reduplicand}$ is used for this category);
- image (aka copy) = the reproduction of the reduplicand (the variable A_{image} is used for this category).

In addition, two further notions are important for better understanding the workings of reduplication, namely

- original meaning-bearing unit = the semantically equipped construction of which the reduplicand forms part;
- reduplicated meaning-bearing unit = the semantically equipped construction which involves both the reduplicand and the image.

I illustrate the above notions with an example from Modern Hebrew (Afro-asiatic) in (1).[5]

4 Mel'čuk (1996) is the source for a number of further important notions which will be referred to in some detail in Section 3.3.

5 Throughout this paper, (supposedly) reduplicative constructions are highlighted in bold in the examples (including the morpheme glosses and the English translations) when they cover only part of a sentence or a word-form. If a reduplicative construction also involves non-reduplicated material, this is indicated by single underlining. Except otherwise stated, the morpheme glosses and the translations are those which are provided in my sources.

(1) Modern Hebrew [Levkovych 2007: 122][6]

a. singleton adjective[7]

zot	*hajeta*	*'uga*	*metuk-a*	
DEM.PROX:F	be:PRET.3SG.F	cake:F	sweet-F	

'This was a sweet cake'

b. reduplicated adjective

zot	*hajeta*	*'uga*	***metuk-a***	***metuk-a***
DEM.PROX:F	be:PRET.3SG.F	cake:F	**sweet-F**	**sweet-F**

'This was a **very sweet** cake'

The singleton adjective *metuka* 'sweet' (hosting the feminine gender agreement marker *-a*) is the domain which is affected by reduplication. The adjective is a bona fide meaning-bearing unit which corresponds to a full-blown syntactic word. Its chain of segments serves as the reduplicand in the copying process which yields the image *metuka*. The image is identical to the reduplicand as to the quality, quantity, and sequence of the phonological segments and morphemes. The reduplicated meaning-bearing unit *metuka metuka* 'very sweet' is semantically clearly different from (though related to) the singleton adjective.

Diagram 1[8] opposes the reduplicand and the image of example (1). Under the heading expression I identify the segments which form the phonological chains of the two units under inspection. To the right of the dividing line, the semantic features are specified for the same units. Boldface and grey shading are used for those cells which host identical fillers for both reduplicand and image. Small caps are used for the labels of grammatical categories.

Unit	Expression						Content	
$A_{reduplicand}$	*m*	*e*	*t*	*u*	*k*	*a*	sweet	FEMININE
A'_{image}	*m*	*e*	*t*	*u*	*k*	*a*	sweet	FEMININE

Diagram 1: Reduplicand-image comparison for (1).

6 The English translations are mine.

7 Example (1a) has been (re-)constructed on the basis of the attested example (1b). The acceptability of (1a) has been confirmed by Nataliya Levkovych (p.c.).

8 Except otherwise stated, in diagrams of this kind, I employ the conventions of the source from which have taken the example. This means that the phonological chain of segments is represented either in the (official) Latinate orthography of a given language or according to the transcription system the author of the source adheres to.

The situation is straightforward since for each of the cells of the reduplicand I find identical fillers in the corresponding cells of the image. There can be no doubt that reduplicand and image are formally and functionally perfect replicas of each other.

In Diagram 2, I apply the above analytical categories to the reduplicand-image pair of (1).

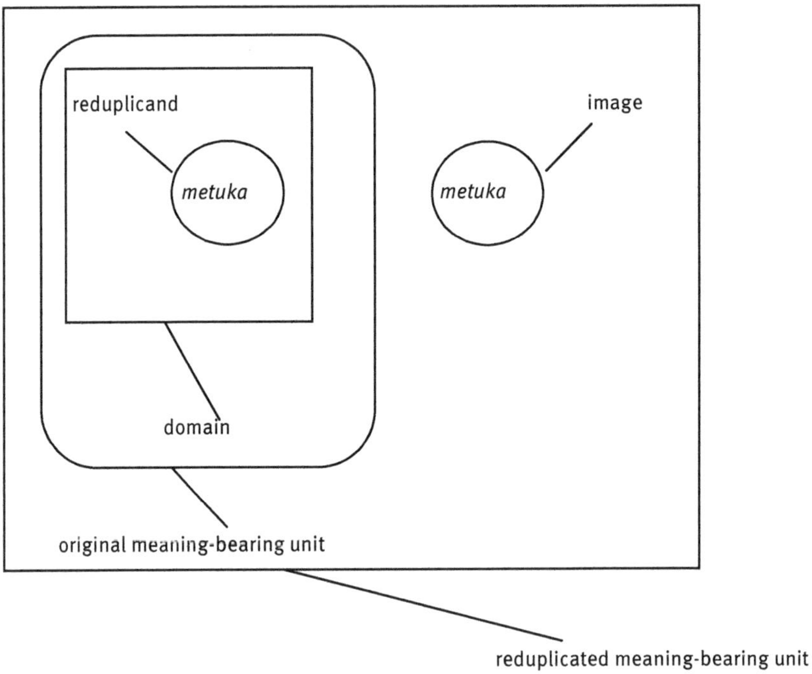

Diagram 2: Basic analytical notions exemplified.

The Modern Hebrew example gives evidence of the identity of three categories. The original meaning-bearing unit corresponds exactly to that of the domain which in turn is coextensive with the reduplicand. A combination of facts of this kind yields an instance of total reduplication of full-blown syntactic words. It is exactly this case which has the highest relevance for the concept of canonical reduplication to which I return in Section 3.2. Before I address the issue of canonicity it is necessary, however, to review some of the problems which arise when reduplication is studied in the absence of an independent point of reference.

2.2 Putative reduplication – some empirical back-up from Oriya and Estonian

2.2.1 Two unequal cases

Why it is necessary to sort things out in the domain of reduplication can be gathered from the subsequent discussion of examples (2)–(3) from two genetically, typologically, and areally unconnected languages. There is ample opportunity to discuss further empirical data along the way (especially through Section 4).

(2) Oriya (Indo-European, Asia) [Neukom & Patnaik 2003: 66][9]

rasta-rɔ	***kuṛ-e***	***kuṛ-e***	*gɔchɔ-guṛie*	*ɔch-i*
road-GEN	**side-LOC**	**side-LOC**	tree-PL	be-3SG

'There are trees **along** the road'

(3) Estonian (Uralic, Europe) [Erelt 1997: 35][10]

oh-ki-ma	*ja*	*puh-ki-ma*
sigh-FREQ-INF	and	puff-FREQ-INF

'to huff and puff'

Both the Oriya and the Estonian example fall under the rubric of potential reduplication because they are in conflict with at least some of the more restrictive definitions of reduplication whereas they are perfectly fine for approaches like that of Maas (2007: 2–5).[11] Since in both (2) and (3) the configurations transcend the word-boundary, they would be automatically counted out, however, by Mattes (2014: 34) because they are considered to be instances of "[s]yntactic repetition" and therefore are only "superficially similar to reduplication".[12] For Stolz (2007a), on the other hand, the Estonian example (3) fails to meet the criteria for being classified as an instance of reduplication because the two sides of

9 In the grammar from which I have drawn the Oriya examples, one of the constituents of a construction of syntactic reduplication is simply glossed RDP (= reduplication) without further analysis of the morphological make-up of the word-form. To demonstrate that the members of the reduplicative construction under inspection are formally absolutely identical, I deviate from the practice of my source by way of providing morpheme glosses for both constituents.

10 The morpheme glosses and the segmentation into morphemes are mine.

11 This approach continues the tradition of holistic conceptions of reduplication as initiated by Pott (1862), Brandstetter (1917), and to some extent also by Moravcsik (1978).

12 This judgment is based on the definition of reduplication put forward by Hurch & Mattes (2007, 2009) who – like many other experts in this field – consider reduplication to be a strictly word-morphological phenomenon.

the linguistic sign are affected only to some extent, if at all. There is thus a competition of at times widely different definitions of reduplication such that it is unclear whether linguists refer to the same concept when they make statements about the properties and the distribution of reduplication across languages.

According to the principles formulated by Stolz et al. (2011), sentence (2) from Oriya contains a bona fide example of total reduplication:

- There are two strings of segments which are identical to each other
  - on the phonological level (both are disyllabic and involve the four segments /k/, /u/, /ɽ/, and /e/ in exactly the same order),
  - on the morphological level (both are bimorphemic and fully functional syntactic words hosting the regular locative suffix *-e*), and also
  - on the semantic level (if taken in isolation separately each of the syntactic words can be translated into English as 'at/on the road').
- They are direct neighbors on the syntagmatic axis.
- Only in this binary combination do they express the meaning 'along the road' which cannot be expressed by a singleton *kuṛe* 'at/on the road' alone.

This situation of formal and functional equivalence is visualized in Diagram 3 for which I use the same conventions as in the case of Diagram 1.

Unit	Expression				Content	
$A_{reduplicand}$	*k*	*u*	*ṛ*	*e*	road	LOCATIVE
A'_{image}	*k*	*u*	*ṛ*	*e*	road	LOCATIVE

Diagram 3: Reduplicand-image comparison for (3).

The identity of reduplicand and image is perfect.

In point of fact, *kuṛe kuṛe* 'along the road' is an instantiation of a more general construction frame $[N_x\text{-LOC } N_x\text{-LOC}]_{prolative}$ (with $N_x = N_x$) which is systematically employed in Oriya to express the prolative[13], i.e., we are dealing with a highly grammaticalized construction with a meaning of its own which does not correspond to the added meanings of its constituent parts. The two components of this asyndetic construction resemble each other to the extent that it is difficult to determine which of the two functions is the reduplicand and which is the image. A and A' cannot be told apart solely on the basis of the above data. What

13 Neukom & Patnaik (2003: 65) subsume the prolative under the umbrella of distributivity.

is clear in any case is that the Oriya example (2) is of the same kind as our initial example (1) from Modern Hebrew in the sense that there is identity of the original meaning-bearing unit, domain, and reduplicand, i.e. total reduplication of syntactic words applies.

In contrast to the Oriya case, the relation of the example (3) from Estonian to the notion of reduplication is by far less straightforward. In the absence of a sentential context, I have to rely exclusively on the explanations provided by Erelt (1997: 34–35) who classifies (3) as an instance of

> [s]imilarity reduplication [which] is a phenomenon, where a formally and/or semantically but not an identical word stem is added (by means of a connective conjunction or without it) [...]. The condition of poeticalness is similarity, i.e. variation where things otherwise coincide. As far as reduplication is concerned, one has to do with stem variation both in the sense of form and meaning. The coincidence provides quantity + emotivity, variation will add poeticalness.

According to this quote the Estonian example belongs to the realm of style whereas the Oriya example is associated with grammar. What further distinguishes (2) from (3) are the following properties of the latter:
- In the Estonian case, there is a conjunction which connects the two infinitives to each other (= syndesis),
- the combination of the two infinitives does not give rise to a meaning that is different from the sum of the meanings of the two conjuncts,
- there is no identity on the segmental phonological level,
- the coordination of the two infinitives is not compulsory for the expression of any function.

Erelt (1997: 35) interprets (3) as an instance of variation exclusively of form and thus implicitly assumes that the content of the two conjuncts is the same so that one might speak of reduplication on the semantic level. To better explain why this is problematic in the first place, I check in what way the members of the Estonian case resemble each other. The two lexical components of (3) are contrasted with each other in Diagram 4.[14]

14 The semantic features I assume are based on the semantic description of the verbs *ohkima* and *puhkima* in the monolingual Estonian dictionary compiled by Langemets et al. (2009). I am grateful to Aina Urdze for her help with placing and interpreting this important tool.

Expression							Content				
	o	*h*	*k*	*i*	*m*	*a*	breathe	FREQUENTATIVE	audible	in sorrow	long
p	*u*	*h*	*k*	*i*	*m*	*a*	breathe	(FREQUENTATIVE)	audible	heavy	short

Diagram 4: Resemblance between the two content words in (3).

Dark grey shading highlights those cells which are identically filled for the two content words in the Estonian example. Light grey coloring is used for a doubtful case to be addressed below. Both of the phonological chains consist of three syllables the last two of which are segmentally identical so that the words rhyme with each other. The initial syllables are closed and display the coda-consonant /h/. Of altogether thirteen segments, ten are the same for both word-forms not only in terms of their quality but also as to the order they come in. Morphologically there is similarity too since both of the word-forms host the derivational morpheme *-ki* which forms frequentative verbs (Viitso 2003: 71) and the word-final infinitive marker *-ma*. There is thus morphological correspondence of the two word-forms to each other. On the content side, we find that the audible activity of breathing forms part of the semantic features of both verbs.

Superficially, it does not seem too far-fetched to treat the Estonian case (3) as a relative of the examples from Modern Hebrew and Oriya in (1)–(2). However, on closer inspection, it is evident that one cannot sweepingly lump all these cases together. First of all, in Diagram 4, the similarities for the word-forms are shown to be limited since several cells escape being highlighted in grey. The root morphemes of the two infinitives only overlap as to their consonantal coda. One of the syllables (= /oh/) is naked in the sense that it lacks a consonantal filler of the onset position whereas its counter-part boasts a filled onset (= /puh). The nuclei of the syllables are qualitatively different (mid-high back /o/ ≠ high back /u/). Since the differences arise exactly with those morphemes which carry the lexical meaning whereas the similarities on the expression side are restricted to the bound morphology, we are dealing with a morphological minimal-pair.

Maas (2005: 398) distinguishes lexical reduplication and grammatical reduplication. The former involves cases which involve "the repetition of word-internal lexical elements in a sentence" whereas grammatical reduplication is said to apply when lexical morphemes "derive from different stems while the affixes are identical (in function, at least)". This situation is basically the same as morphological agreement of syntactic words. If it is possible at all to connect agreement to reduplication, it is certainly the case that the examples from Modern Hebrew in (1) and Oriya in (2) involve the identity not only of the lexical morphemes but also of the affixal morphemes. In contrast, the Estonian example (3) has more in common with Maas's category of grammatical reduplication, if at all.

Semantically, *ohkima* 'to sigh' and *puhkima* 'to puff' do not share all of their features. The verbs describe clearly different pulmonically supported activities of different duration and intensity. The overt marking of the verbs for the frequentative notwithstanding, only *ohkima* is characterized semantically as describing a repeated action. As to the meaning of *puhkima*, however, repetition is not mentioned explicitly as a semantic feature. In connection to this issue, Aina Urdze (p.c.) draws to our attention the fact that the frequentative *ohkima* 'to sigh repeatedly' contrasts with the formally factitive verb *ohkama* 'to sigh' which is semantically clearly related. The possibility to identify a semantic relation between *puhkima* 'to puff' and *puhkama* 'to rest' is barred. In the absence of a corresponding semelfactive verb, *puhkima* thus neutralizes the semelfactive and frequentative distinction by way of functioning as a general verbal expression of puffing. Therefore, the morphological identity of the affixal parts of the two verbs does not yield full identity on the content level. It is appropriate to understand their coordination as a kind of bilateral semantic completion in the sense that the two verbs jointly cover all aspects of the activity of a given participant. In isolation, neither of the verbs is semantically sufficient to describe the same breathing event.

The combination of the two infinitives in (3) is in line with the broad view on reduplication taken by the proponents of MDT. Inkelas & Zoll (2005: 61) invoke a "family of construction types" to which reduplication belongs.[15] This family consists of six members which differ from each other as to the extent to which the morpho-semantic features of the constituents of the constructions are in agreement with each other. Table 1 reproduces MDT's "cline of semantic similarity" (Inkelas & Zoll 2005: 62) verbatim. The very idea of a cline as such has an impact on our concept of canonical reduplication as will become evident from Sections 3.2–3.3 below.

15 For a detailed criticism of this approach, the reader is referred to Wälchli (2007: 101–103).

Table 1: MDT's cline of semantic similarity.

Type	Features
reduplication	all features agree
synonym constructions	all features agree except stratum/register/etc.
near-synonym constructions	most features agree
members of the same semantic set	basic semantic category features agree
semantic inclusion	the features of one are a proper subset of the features of the other
antonyms	the two are specified with opposite values for certain features

Grey shading shows where on the supposed cline the Estonian example (3) might be located. What we see immediately is that the Estonian case is relatively far removed from the top-ranking reduplication. This topmost position, however, is where the examples (1)–(2) from Modern Hebrew and Oriya can be placed.

The meanings of *ohkima* 'to sigh repeatedly' and *puhkima* 'to puff' are at best only partly synonymous. They are certainly related in the sense of belonging to the same semantic sphere. However, I doubt that it is possible to replace the one with the other in all contexts. In the case of (3), it is inappropriate to categorize the example as a full-blown construction – that is why I speak of a configuration. More precisely, what we are dealing with is a kind of stylistically exploited collocation. Moreover, (3) is an instantiation of the general pattern of Estonian binary coordination which is [X *ja* Y]$_{\text{coordination}}$. This means that example (3) cannot be told apart from any other kind of coordination on structural grounds. Put differently, Erelt's similarity reduplication is largely indistinct from other constructions which have nothing at all to do with reduplication. These factors render it difficult to apply to the Estonian example (3) the analytic notions (such as that of the original meaning-bearing unit) I have introduced in Section 2.1.

2.2.2 Towards the continuum

The two languages under inspection in this section display further configurations which pass as instances of reduplication according to the descriptive-linguistic sources I refer to. For Oriya, cases like (4)–(5) are reported.

(4) Oriya [Neukom & Patnaik 2003: 65]

e-sɔbu	*jinisɔ*	***gãã***	***gãã-re***	*sɔhɔjɔ-re*	*miḷ-uch-i*
this-all	thing	**village**	**village-LOC**	easy-LOC	be_available-PROG-3SG

'All these things are easily available **in any village**'

The formal difference between the configurations in (2) and (4) is the presence/absence of the case marker on the first member. The absence of a case morpheme in (4) can be interpreted in two ways: either the first member of the configuration appears in the bare stem form or it is in the unmarked nominative. In (2), both members are inflected overtly for the locative whereas there is a kind of group-inflection in (4) where the locative suffix is attached only to the second member. As far as I can tell on the relatively small empirical basis I have access to, the omission of the locative marker helps to distinguish the prolative construction $[N_x\text{-LOC}\ N_x\text{-LOC}]_{prolative}$ from the distributive construction $[N_x\ N_x\text{-CASE}]_{distributive}$.[16]

In Diagram 5, I contrast reduplicand and image according to the principles employed in connection with the previous diagrams. To distinguish reduplicand and image for the purpose of Diagram 5, I have stipulated a sequence reduplicand > image. The rightward linearization serves as default option for this and the subsequent cases discussed in this section.

Unit	**Expression**				**Content**	
$A_{reduplicand}$	*g*	*ã:*			village	(NOMINATIVE)
A'_{image}	*g*	*ã:*	*r*	*e*	village	LOCATIVE

Diagram 5: Reduplicand-image comparison for (4).

As can be gathered from Diagram 5, reduplicand and image are not absolutely identical phonologically, morphologically, and semantically. The stem which conveys the lexical meaning is the same for both reduplicand and image. However, the latter host an additional case suffix that has no equivalent on the reduplicand so that only the image is morphologically complex (= dimorphemic). The presence of the case suffix is also responsible for the higher degree of phonological complexity of the image in contrast to the reduplicand. The reduplicand is monosyllabic and counts only two segments whereas the image is disyllabic with altogether four segments. On the semantic level, the reduplicand and the image

16 Neukom & Patnaik (2003: 66) cautiously conclude that "the case suffix can be repeated or not."

have only one feature in common. The locative is an exclusive feature of the image. The question arises whether it is correct to distinguish reduplicand and image in exactly this way for (4) in the first place. Would it make more sense alternatively to apply an analysis which assumes root reduplication or stem reduplication prior to inflection? I will return to this problem in Section 4.1 where I elaborate on the reduplication of roots and stems.

In addition to (4), my source mentions at least two further types of potential reduplication which are connected to the notion of reflexivity as shown in (5).

(5)		Oriya				[Neukom & Patnaik 2003: 108]
	a.	*Rɔbi*	***nijɔ-<u>ku</u>***	***nij-e***	*gaḷi*	*de-l-a*
		Rabi	**REFL-<u>DAT</u>**	**REFL-NOM**	scolding	give-PST-3SG
		‘Rabi scolded **himself’**				
	b.	*Sita*	***nijɔ***	*upɔr-e*	***nij-e***	*hɔs-il-a*
		Sītā	**REFL.OBL**	top-LOC	**REFL-NOM**	laugh-PST-3SG
		‘Sītā laughed **at herself’**				

Functionally, the configurations in (5a–b) are said to be identical – they both serve the purpose of emphasizing the reflexive component (Neukom & Patnaik 2003: 108). Emphasis is not a properly grammatical function. It belongs to the domain of pragmatics, because no change of meaning is associated with this kind of configuration. This is the reason why (5a–b) are better classified as instances of repetition.[17] If we discount this general problem for the sake of the argument, we can identify further properties which distinguish (5a–b) from the prior examples from Oriya. In (5a), both members of the configuration host case suffixes. However, the leftmost member is inflected for the dative whereas the rightmost member displays nominative morphology. There is thus neither phonological nor morphological identity of the two members. The morphological dissimilarity may be interpreted as evidence of their different syntactic status. Neukom & Patnaik (2003: 49) state that “[t]he nominative case is used to mark the subject of a sentence” whereas “[t]he dative case suffix *-ku* is used to mark [...] occasionally the patient of transitive verbs (object)” (Neukom & Patnaik 2003: 50). It is therefore unquestionable that in (5a) [*nijɔ*$_{\mathit{REFL}}$-*ku*$_{\mathit{DAT}}$]$_{\text{patient=object}}$ and [*nij*$_{\mathit{REFL}}$-*e*$_{\mathit{NOM}}$]$_{\text{agent=subject}}$ are part of different constituents of the sentence. The dative identifies the reflexive pronoun as internal argument, i.e. as part of the VP,

17 For reasons of space, I do not go into the intricacies of determining the dividing line which separates repetition from reduplication. To summarize the results of Stolz & Levkovych (accepted), one may say simplifying that under repetition something is said twice without any change of meaning whereas in the case of reduplication a new meaning or function arises.

while the nominative case identifies the reflexive pronoun as external argument which is located outside the VP.

In the case of (5b), the two members of the configuration are not only morphologically distinct from each other but are also separated from each other by an intercalated relational noun *upɔre* 'on top' as postposition which governs the oblique case on its pronominal complement (Neukom & Patnaik 2003: 327–328). In all other cases of Oriyan potential reduplication, the members of the configuration are neighbors of each other in the sense of directly adjacent syntactic words. The two instances of the reflexive pronoun belong again to two different syntactic constituents. As in (5a), the nominative *nije* is the external argument in (5b). In contrast, the postpositional phrase *nijɔ upɔre* 'at herself' is embedded in the VP. In both (5a) and (5b), the members of the configuration are situated on different sides of the major syntactic divide of the sentence. In the case of (1) and (2), however, no comparable boundary runs between the reduplicand and the image. Thus, one might want to argue that the reduplicand-image analysis cannot be applied to cases like (5a–b) at all.

For Estonian, Erelt (1997) mentions an array of further candidates for the status of potential reduplication from which I select only three to have an equal number of types for the two languages under discussion. First of all, there are also examples of total reduplication as in (6).[18]

(6) Estonian [Erelt 1997: 14][19]

kohe-kohe	*algab*	*seanss*
soon-soon	start:3SG	show

'[...] the show is going to begin **very soon**'

The segmental chain of the adverb *kohe* 'soon' occurs twice in example (6). The two identical chains of segments are adjacent to each other. Orthographically they are considered to yield one word (= compound). The function of the pattern $[\text{ADJ}_x/\text{ADV}_x\text{-ADJ}_x/\text{ADV}_x]_{\text{intensive}}$ is the intensification of the meaning of the singleton syntactic word. It is possible therefore to speak of a construction. In this case, the original meaning-bearing unit is the same as the domain which is identical with the reduplicand. We therefore have another parallel to the Modern Hebrew example (1) with the proviso that the Estonian *kohe-kohe* 'very soon' does not involve

18 In Erelt's (1997: 13) terminology, we are dealing with "the main type of non-bounded reduplication [which] is total asyndetic reduplication."

19 The morpheme glosses are mine. I have also left out the first part of Erelt's original example because it has no connection to the sentence structure of example (6) in the first place.

any inflectional morphology. Diagram 6 shows that the Estonian example (6) yields the highest possible score as to the identity of reduplicand and image.

Unit	Expression				Content
$A_{reduplicand}$	*k*	*o*	*h*	*e*	soon
A'_{image}	*k*	*o*	*h*	*e*	soon

Diagram 6: Reduplicand-image comparison for (6).

There are undeniable similarities of the Estonian case (6) and the Oriya example (2). Some parallels can also be detected when we compare the Estonian example (7) to the Oriya examples (4) and (5a).

(7) Estonian [Erelt 1997: 32][20]
Seda *tuleb* *teha* ***järk~järg-u-lt,*** *mitte* *korraga*
this:PTV come:3SG do:INFII **step~step-PL-ABL** NEG at_once
'It has to be done **step by step**, not all at once'

The noun *järk* 'step' is represented twice in (7). It appears in its zero-marked base form (= nominative or stem) *järk* to which the regular ablative plural *järgult* 'from steps' of the same noun is added. The voicing of stem-final /k/ to /g/ in combination with the plural-marker *-u* is a regular morphonological process. Reduplicand and image are phonologically similar only as to the shared stem. The two word-forms from the same paradigm are considered to be an orthographic unit. Since the nominative bears no case marker, one might want to equate this configuration formally to (4) from Oriya so that both are instantiations of [$N_{x(stem)}$ N_x-CASE] or [$N_{x(nominative)}$ N_x-CASE] (though with different functions).

Diagram 7 shows that the correspondences between reduplicand and image are relatively scarce for example (7). Light grey shading is employed for the regular voicing of the stem-final velar plosive.

Unit	Expression							Content		
$A_{reduplicand}$	*j*	*ä*	*r*	*k*				step	SINGULAR	(NOMINATIVE)
A'_{image}	*j*	*ä*	*r*	*g*	*u*	*l*	*t*	step	PLURAL	ABLATIVE

Diagram 7: Reduplicand-image comparison for (7).

20 The morpheme glosses are mine.

Independent of the interpretation of *järk*, the Estonian example (7) is certainly not as good a representative of reduplication as example (6) from the same language.

The classification of (8) as an instance of reduplication is even more problematic.

(8) Estonian [Erelt 1997: 18][21]

Taevas	*on*	***sini<u>semast</u>***	***sinine***
sky	be.3SG	**blue:<u>COMP:ELA</u>**	**blue**

'The sky is **bluer than blue**'

The color adjective *sinine* 'blue' occurs twice in (8), viz. sentence-finally in the positive form *sinine* 'blue' and in the slot immediately to the left in the elative of the comparative as *sinisemast* '(literally) of the bluer'. Erelt (1997: 15) claims that "comparative reduplication is a rather common type of intensifying reduplication." However, what my source does not state explicitly is the fact that comparative reduplication (= [ADJ_X-COMP-ELA ADJ_X(-COMP)]) resembles closely one of the general comparative constructions of Estonian which comes in the shape of $[\text{N-ELA ADJ-COMP}]_{\text{comparative}}$.[22] The configuration therefore is only minimally distinct from functionally related non-reduplicative constructions. Furthermore, not only are the two word-forms of the adjective morphologically distinct, but there also runs a semanto-syntactic border between *sinisemast* (which functions as expression of the standard of comparison) and *sinine* (which identifies the quality) in (8). Syntactically, *sinisemast* is an attribute to *sinine* which serves as its head. This head-modifier distinction suggests that we are dealing with two syntactic words which behave differently from the reduplicand-image pairs in (1), (2), and (6). To some extent, the Estonian example (8) resembles the examples (5a–b) from Oriya which involve configurations whose members belong to different syntactic constituents.

Intuitively, the Oriya case in (2) and its Estonian equivalent in (6) are much better representatives of reduplication than all the other cases discussed above. This means that (2) and (6) are acceptable as instances of reduplication without much explanation whereas it is necessary to explain explicitly why (3), (4), (5a–b), (7), and (8) also count as examples of reduplication. These doubts notwithstanding, I try to do justice to all of the above examples no matter how they

---

21 The morpheme glosses are mine.

22 As in (i)

(i) Estonian [Lavotha 1973: 94]

Ta	*on*	*minust*	*noorem*
s/he	be.3SG	1SG:ELA	young:COMP

'S/he is younger than me'

would fare under the application of restrictive definitions of reduplication. Diagram 8 shows that it is possible to arrange the more or less convincing representatives of reduplication on a continuum which takes the degree of the phonological, morphological, syntactic, and semantic (dis)similarity of the constituents of the constructions as one of its criteria the other two being wordhood of the constituents and syntactic adjacency.[23] Grey shading identifies those cases which fail to meet the latter criterion. The placement of the configurations on the continuum is only approximate. The continuum can be employed for the purpose of language comparison as well as for the clarification of language-internal questions.[24]

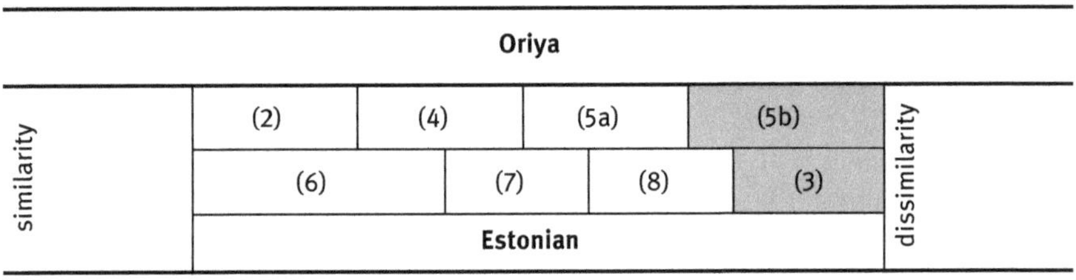

Diagram 8: The continuum – preliminary version.

What we see is that the configurations of the two languages cover much the same space between the two extremes of the continuum. This does not imply that the Oriyan and Estonian configurations occupy exactly the same sections on the continuum. The examples (2), (4), and (6) display the highest degree of similarity of the members of the configurations whereas similarity diminishes the further to the right we move on the continuum. The cases of (3), (5a–b), (7),

23 The scores on which the continuum in Diagram 8 is based are as follows:
- Examples (2) and (6) are closest to the pole of similarity because the two members of the configuration are identical on all levels from phonology to semantics, taken in isolation each of them displays the appropriate form of a syntactic word, and the constituents are direct neighbors syntagmatically.
- Example (4) fulfils most of the criteria except for those of morphological identity and separate wordhood.
- Examples (5a), (7), and (8) fail to meet the criteria of morphological and semantic identity. Their members are not of equal status syntactically.
- As to examples (5b) and (3), several of the criteria are violated against. In both cases, there is no syntactic adjacency of the syntactic words under consideration. In the case of (3), the stems of the coordinated infinitives are different so that dissimilarity applies on the levels of phonology, morphology, and semantics.

24 The bracketed numbers in Diagram 8 refer back to the examples from Oriya and Estonian.

and (8) involve an increasingly more important aspect of internal dissimilarity. We learn from the provisional continuum that it is feasible to order the phenomenology of reduplication in a linguistically meaningful way. It is the main task of this paper to prove that searching for the principles of order makes perfect sense for the research program that is dedicated to reduplication in general.

The variegated picture of the above cases notwithstanding, I do not want to exclude the possibility that there might be a common denominator which connects all of them. This does by no means imply that all of the cases are equally matched representatives of the same category; it might turn out that there are good reasons to exclude some of the candidates from the membership of the category of reduplication. I assume that it is not helpful for refining my understanding of the important role reduplication plays in human language in general to approach the domain holistically but at the same time in an ill thought out manner – like Pott (1862), the pioneer of linguistic investigations on reduplication did in the mid-19th century. If one treats all the phenomena on a par with each other, one might also miss the opportunity of discovering the internal architecture of the domain. To avoid these pitfalls, it is necessary to investigate reduplication on the basis of a language-independent fixed reference point which additionally allows us to explore and eventually also to determine the outer limits of the category under scrutiny. To this end, the concept of canonical reduplication provides a suitable reference point which is especially useful to refine the as yet rather crude version of the above continuum. This paper is meant to introduce canonical reduplication and identify its (largely, but not exclusively methodological) advantages.

3 From prototypical reduplication to canonical reduplication

3.1 Towards prototypicality

Inkelas & Zoll (2005: 61) assume that "in prototypical reduplication, the daughters agree in every morpho-syntactic feature" (cf. Table 1 above). The notion of the prototype is thus invoked but not spelled out in every possible detail. The quoted authors are preoccupied mainly with sketching the above-mentioned family of construction types most of which do not even fall under the rubric of reduplication in the framework of MDT. What, however, has gone missing so far is the application of the prototype-model to the wide variety of phenomena whose admission to the category of reduplication is uncontroversial. The central axiom of MDT is that each and every reduplicative construction is underlyingly patterned according to the same model whose surface realization requires a succession of intermediate steps which affect mostly the phonology of the constituents of the construction (Inkelas & Zoll 2005: 6–20). The reliance on morpho-semantically identical underlying forms may be a major obstacle for recognizing that reduplicative constructions are not generally all of the same kind. I take issue with the undifferentiated view of reduplication.

In the same year as the above MDT-borne proposal, Wälchli (2005: 166) provides a diagram which "lists partial and full reduplication together with some less prototypical types of reduplication, aligned on a tight-loose reduplication scale." Not only is the notion of prototypicality invoked in this quote but Wälchli's scale can also be understood as a kind of continuum which is different from the one I have presented in Diagram 8 above. My adaptation of Wälchli's scale is given in Diagram 9.

repetition of sentences	repetition of phrases	repetition of words (without connective prosody)	full (stem) reduplication (word iteration)	partial reduplication (affixal reduplication)	gemination of vowels or consonants

loose reduplication ◄- ► tight reduplication

Diagram 9: Wälchli's (2005: 166) scale of different types of reduplication.

Grey shading[25] highlights those types of reduplication which Wälchli (2005: 166) considers to be prototypical. The above scale serves to support Wälchli's (2005: 167) argument according to which

> full (but not partial) reduplication is formally related to co-compounds and has some functions similar to some co-compounds (as in distributive contexts). There are, however, some important differences between full reduplication and co-compounds.

It is Wälchli's aim to contour the domain of co-compounds by way of determining the borderline that separates them from proper reduplication. That is why the scale does not name further sub-divisions of reduplication explicitly. Total reduplication and partial reduplication seem to be treated as being equally

25 In the text from which I reproduce the scale, the cells occupied by full reduplication and partial reduplication are marked graphically as special by extra-thick frames.

prototypical. Since the criterion for the placement of the phenonema on the scale is increasing/decreasing looseness/tightness of the construction, the prototypical cases are not located on the extremes of the scale. On the one hand, Wälchli's scale is suggestive of the possibility to rank reduplicative constructions according to the properties they display. On the other hand, the scale is still too undifferentiated to allow us to assign each instance of potential reduplication its appropriate place on the continuum. In what follows, I will demonstrate that it is indeed possible to refine the scale in terms of prototypicality and canonicity.

In a recent paper, Stolz et al. (2015: 826) provide a list of altogether eight criteria which are meant to characterize the prototype of reduplication.[26] According to these criteria, a reduplicative construction can be considered to be in accordance with the prototype if it combines the following properties:

i. exactly one linguistic sign is involved (as domain),
ii. the construction is strictly binary,
iii. absolute phonological identity of the constituents applies in quantity and
iv. quality of their segments,
v. the constituents are syntagmatically adjacent,
vi. without rendering any other structural unit discontinuous,
vii. the construction is different from the singleton units in terms of their semantics and/or functions,
viii. the construction has the status of a syntactic word.

Any failure to fulfill one or several of the above requirements removes a given construction from the prototype. Since deviations from the prototype may cumulate on several of the parameters, different constructions might wind up at different distances from the prototype such that a radial structure emerges with its center being occupied by the prototype whereas those constructions which are least in line with the above criteria are situated on the margins (Lakoff 1987: 91–114). The continuum in Diagram 8, for instance, could be reorganized radially by way of counting how many of the above criteria are met by the individual constructions. Table 2 indicates how the constructions covered in Diagram 8

26 This list reflects the groundbreaking thoughts of Mel'čuk's (1996: 43–44) and updates the earlier proposal of a prototype put forward in Stolz et al. (2011: 42). The notion of prototypical reduplication has been put to the test successfully in recent work by Nintemann (2016: 5–8), Otsuka (2016: 24–32), and Robbers (2016: 100–103).

fare in terms of prototypicality.[27] Grey shading additionally highlights all instances of conformity to the prototype. There is only one construction which reaches the highest possible score of eight times YES, namely example (6) from Estonian. At the same time, there is also only one criterion which is met by all eight examples, namely criterion (vi.), i.e. none of the constructions interrupts other constructions.

Table 2: Check of prototypicality.

Example	**Criteria/Prototype**							
	i	ii	iii	iv	v	vi	vii	viii
(2)	YES	YES	YES	YES	YES	YES	YES	NO
(3)	?	NO	NO	NO	NO	YES	?	NO
(4)	YES	YES	NO	NO	YES	YES	YES	YES
(5a)	YES	YES	NO	NO	YES	YES	YES	NO
(5b)	YES	YES	NO	NO	NO	YES	YES	NO
(6)	YES	YES	YES	YES	YES	YES	YES	YES
(7)	YES	YES	NO	NO	YES	YES	YES	NO
(8)	YES	YES	NO	NO	YES	YES	YES	NO

In Table 2, the constructions come in the order in which they have been introduced in Section 2.2 above. The matrix lends itself to being transformed into a radial category as shown in Diagram 10.

The Estonian example (6) is at the center of this succession of concentric circles because it is characterized by positive values for each of the parameters of prototypicality established above. The least degree of prototypicality is associated with example (3) which incidentally also comes from Estonian. In stark contrast to (6), (3) receives a YES only on one parameter – and that is criterion (vi.) for which all of the constructions have a YES. What the format of Diagram 10 does not convey adequately is the huge gap between two outlier circles. The gap, however, can be reconstructed from the information given in Table 2. Where example (5b) from Oriya still has a score of four times YES, the score of example (3) is down to just a single YES, i.e. the distance of the latter from the

27 In Table 2, a cell may host a YES if the criterion of prototypicality is met, a NO in case of a violation of a requirement, and the interrogation mark if it is unclear how to interpret the facts.

center is four times as long as that of the former. On this basis, it can be concluded that (3) is a bad representative of reduplication, if at all.

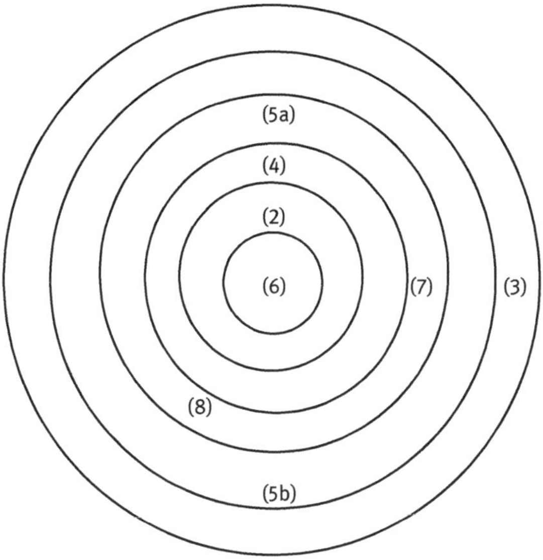

Diagram 10: Radial category – prototype.

Diagram 10 tacitly assumes that the criteria (i–viii) are of equal importance so that there is no ranking order of strength. However, this supposed equality still needs to stand the test. It may be the case that some of the criteria are more important than others. Moreover, the list of parameters calls for being thoroughly revised since it has never been declared to be exhaustive or final. In Section 3.2, I demonstrate that, for the questions raised in research on reduplication, canonicity provides an excellent point of departure which has a number of advantages over prototypicality.

3.2 Canonicity in general

In prototype-based approaches, "the best example of a category" (Lakoff 1987: 24) is a central notion because they provide the model for the entire category. The best example is not an abstraction but a realized life-form of a given category. For the issue under debate, this means that the prototype comes in the shape of an attested construction. The continuum in Diagram 8 and the radial category in Diagram 10 are strongly suggestive of total reduplication being the prototype of reduplication. However, contrary to the expectations e.g. of Rubino (2005a), languages may have reduplication without giving evidence of the prototype (Stolz et al. 2015: 806–808). For the indigenous languages of North America and Central America, Robbers (2016: 127–151 and 195) shows that 40 % of her sample of a hundred languages attest exclusively to partial reduplication, i.e. the supposedly prototypical total reduplication is absent from a sizable group of languages.[28] Not only is this result unpredictable on the basis of Rubino's assumption that the presence of partial reduplication implies the presence of total reduplication (but not vice versa) but it is also indicative of the limitations of the prototype approach in the domain of reduplication.

In contrast to the prototype approach, the canonical approach[29] as advocated by Corbett (2005: 25)

> sidesteps two potential dangers in typology, namely 'premature statistics' and 'not comparing like with like.' The first danger is that something which is frequently found may be treated as uninteresting, whereas there are linguistic phenomena which are common yet which [...] should surprise us. The second danger is that we fail to take sufficient care over our terminology and so do not see that phenomena labeled identically are in fact distinct (conversely we miss identities because of different traditions of labeling).

These are exactly my thoughts as to the research traditions in the domain of reduplication. Thus, the canonical approach seems to be tailor-made for solving the most pressing issues of mine. Corbett (2005: 26) describes the canonical approach as follows:

> In a canonical approach, we take definitions to their logical end point and build theoretical spaces of possibilities. Only then do we ask how this space is populated. [...] It follows that canonical instances (the best examples, those most closely matching the canon) may well not be the most frequent. **They may indeed be extremely rare, or even non-existent** [boldface added]. However, they fix a point from which occurring phenomena can be calibrated, and it is then significant and interesting to investigate frequency distributions.

---

28 In Nintemann's (2016) necessarily much smaller sample of twenty languages of Kenya, some of the Cushitic languages – most notably Dahalo and Oromo – attest exclusively to partial reduplication. Thus, the absence of the prototype cannot be demoted to the status of an areal feature of the Americas.

29 In the unlikely event that the concept of canonicity is misunderstood, I hasten to clarify that the canonical approach does not make statements about correct vs. incorrect, right and wrong, etc. Canonicity is not prescriptive but only an idealization. It is therefore different from the concept of linguistic naturalness because the latter assumes that language structures are programmed inherently to develop towards more naturalness when they are given the chance under the conditions of language change (Dressler 1985).

From the sentence I have highlighted in this quote, we learn that canonicity is an abstract notion which may or may not find a correlate in actually attested linguistic phenomena. As a matter of fact, Corbett sketches a kind of yardstick and not necessarily structures which are identifiable in a given language. If the function of canonicity is largely that of a tool of measurement, it is perhaps misleading to equate "canonical instances" with "the best examples, those most closely matching the canon", because if the instances do not replicate the canonical pattern to a hundred percent they are at best "relatively" canonical. They are best examples only in the sense that there are no competing examples which fulfill the requirements of canonicity to a higher degree.

The canonical approach has the advantage that the absence of empirical evidence of canonical instances in a given language poses no problem at all. The data from any number of languages can be investigated under language-independent fixed laboratory conditions, in a manner of speaking. Moreover, the notion of canonicity provides a suitable (mostly terminological) escape route for all those linguists who in the light of Haspelmath's (2006) deconstruction of the notoriously ill-defined concept of markedness no longer feel comfortable with the marked-unmarked opposition which we have inherited from Praguian structuralism. Note that canonical does not translate unmarked and non-canonical does not translate marked or vice versa. Determining (non-)canonicity does not constitute an evaluation of linguistic facts in terms of their markedness. Whether or not the results which are obtained in the framework of the canonical approach can be interpreted afterwards according to the principles of markedness theory (Bybee 2011) is a completely different story.

For the purpose of this study, we still need to "take definitions to their logical end point" first before we can set out to check the empirical data for canonicity. Since the extant definitions of reduplication (tacitly or not) take attested constructions (or configurations) as their point of reference, it is advisable to abstract from the previous attempts at defining reduplication. According to the philosophy of the canonical approach, the logical end point of the definition of reduplication should be reached via data-independent reasoning in order to avoid shaping inadvertently canonical reduplication on the structural properties of instances of reduplication as attested in individual languages. In the subsequent Section 3.3, I put forward a first version of canonical reduplication.

3.3 Canonical reduplication – as an abstraction

As to the necessity of designing canonical reduplication, one might want to object that the criteria (i–viii) which circumscribe prototypical reduplication in

Section 3.1 already provide a picture of canonicity in the domain of reduplication so that there would be no need for further research in this direction. However, I claim that the above list of criteria is not only insufficient since it does not give full coverage to all of the parameters which are crucial in the case of reduplication, but it also lacks independence from the actually attested data. Starting anew on the basis of the canonical approach allows us to shake off the unwanted interference by our previously acquired and probably skewed knowledge about reduplication in (some of) the languages of the world. In what follows in this section, I cannot help making my point in a somewhat commonsensical style without going deeply enough into the matter at hand. To argue these issues properly, the format of a paper in an edited volume is by far too limited so that I will address the topic of this section again and elaborate thereupon in a separate monograph.

To cut a potentially never-ending discussion short right at the beginning, I stipulate that a given constellation of structural facts counts as canonical reduplication if and only if it fulfills the requirements that define constructions (Stolz 2006).[30] This stipulation has a bearing on several of the points I am about to make in the subsequent paragraphs – especially, but not only, when it comes to talking about the semantics and/or functions of canonical reduplication. With this axiom in the back of my minds, I set out to giving an account of canonical reduplication as an abstract concept.

Going by its etymology, the English[31] term reduplication presupposes that something is doubled[32] in the sense of being copied, duplicated, replicated, or reproduced.[33] For replication, etc. to be the case there must be an original entity

30 The status of construction distances canonical reduplication from what Kallergi (2015: 18–23) calls "reduplication at the syntactic level". In this way, I can skip discussing instances of repetition because constructions either have grammatical or lexical functions whereas repetition belongs to the realm of style and pragmatics (Stolz & Levkovych accepted).

31 My decision to look into the etymology of the English term has the exclusive function of a convenient opener for the subsequent line of argumentation. My choice does by no means suggest that all other terminological traditions of different national philologies, etc. are subordinate to the Anglo-American school(s) of thought. I am certain that canonical reduplication would always look the same no matter from which terminological tradition I start to develop my ideas.

32 Webster's (1994: 1204) indicates Late Latin *reduplicāt(us)* as etymological source of the English verb reduplicate. The Late Latin participle can be decomposed into the prefix *re-* (= iteration) plus the stem of the verb *duplicāre* 'to double' plus the participle ending *-ātus*.

33 Again according to Webster's (1994: 428) none of the nine meanings of the transitive verb to double invokes the notion of something being copied exactly (except when use in musicology to indicate of reduplicating "by means of tone in another part, either at unison or at an octave

A which serves as the model for another entity A' which is a copy of the former.[34] My model is thus indebted to BRCT to the extent that I assume that a reduplicative construction involves two slots one of which receives its filler by way of copying the item which occupies the filler of the other slot. In our case the word-boundary is transcended (Inkelas & Zoll 2005: 4–5). For A to be copied as A', A must exist prior to A'. This means that A precedes A' on the time arrow. Given the linear nature of speech, it can be concluded that the order in which the reduplicand and its image are uttered is that of Formula I.

Formula I: $A_{\text{reduplicand}} > A'_{\text{image}}$

Thus, rightward reduplication can be considered to be canonical (Stolz et al. 2011: 54–57). Note that Mel'čuk's (1996: 46) parameter of "latéralité" is not included in the above list of criteria of prototypical reduplication exactly because the proponents of the prototype already knew that there is ample evidence of leftward reduplication and a plethora of controversial cases in the languages of the world. Thus, their prior knowledge of the empirical facts precludes the formulation of expectations on this parameter. For canonical reduplication, however, it is irrelevant that rightward reduplication cannot be proved to be the sole option.

A' is best recognizable as an image of A if it can be directly compared to the reduplicand. Therefore, the image should be close to the reduplicand in terms of temporal and topological distance. The further away A' is from A in speech, the more their relationship comes to resemble that of anaphor, i.e., A + A' do not form a tight unit with a distinct constructional meaning or function. In this sense, immediate adjacency as in Formula II is canonical whereas a sequence A X A' (with X = intercalation of any size) is non-canonical to the extent that it is doubtful whether the sequence can be classified as reduplication, in the first place.

Formula II: $[A_{\text{reduplicand}}\ A'_{\text{image}}]_{\text{canonical reduplication}}$

According to criterion (v.) above, prototypical reduplication also requires that the reduplicand and its image are syntactic neighbors of each other. The criterion corresponds to Mel'čuk's (1996: 43–44) parameter of contiguity (Stolz et al. 2011: 48–52).

Prototypical reduplication and canonical reduplication also agree as to the basically binary structure of the phenomena under discussion (Stolz et al. 2011:

above or below"). The dictionary tells us that most of the transitive uses of this verb associate with the idea of increasing size or quantity by the factor 2.

34 In the context of this study it is of course clear that what is copied must be a linguistic entity.

57–58).[35] To support the hypothesis that binary constructions are generally preferred over more complex ones (such as triplications) the principles of economy can be invoked. Multiple images are unnecessary if the desired effect can already be achieved with the minimal number of images. The canonicity of binary constructions does not exclude the possibility that, for instance, triplication exists.

One possible way of determining the nature of $A_{\text{reduplicand}}$ and A'_{image} is to start from the assumption that all human languages have a lexicon which contains the basic units to be fed to syntax, in order to build up utterances.[36] We can then take the shortcut of taking for granted that syntactic words are the best candidates for the role of linguistic entity that is copied when canonical reduplication applies. Formula III captures this idea schematically.[37]

Formula III: $[\omega_{\text{reduplicand}}\ \omega'_{\text{image}}]_{\text{canonical reduplication}}$

If fully-blown syntactic words are subject to canonical reduplication, the process of copying ideally yields a syntagm with two identically filled slots. The input of canonical reduplication is a syntactic word and the output of canonical reduplication is two syntactic words. The fact that the combination of $A_{\text{reduplicand}}$ and A'_{image} counts as a syntagm is at odds with the requirement of prototypical reduplication according to which the output is supposed to be a complex word. Many cases of total reduplication, for instance, are controversial as to their status (Wälchli 2007: 101, Kallergi 2015: 3–7). It is often very hard to decide whether we are facing compound-like word units or binary syntagms (Stolz et al. 2011: 102–105).[38] This uncertainty is of no relevance for the canonical approach though.

Images of reduplicands can be of very different kinds. The most direct relation between image and reduplicand is that of exact identity (because partial

---

35 Note that Mel'čuk (1996: 41) too assumes that under reduplication the copying process normally produces exactly one image of a reduplicand.

36 My choice of syntactic words as the starting point of the definition of canonical reduplication is also based on the assumption that minimal utterances should at least contain one syntactic word. Utterances which do not fulfill this minimality condition are certainly exceptional and require especially marked contexts to be acceptable in communication. Single bound morphemes, for instance, can hardly be expected to have utterance status.

37 The symbol ω is employed as index for syntactic words.

38 In the light of Haspelmath's (2011) deconstruction of the (universal) distinction of words and phrases, one might consider my preoccupation with wordhood vs. syntagm to be somewhat anachronistic. I do not intend to argue for or against the possibility of a language-independent definition of either of the categories. What is important for my approach is the difference that arises between prototypical reduplication and canonical reduplication.

and/or inexact images need special "rules" to be recognizable as images of a given reduplicand). Each property that is characteristic of the reduplicand is also found with the image and vice versa (Stolz et al. 2011: 45–48). Neither the one nor the other displays features which are exclusive to only one member of the pair. This means that in addition to being exact the reduplication is required to be complete as well (Stolz et al. 2011: 43–45). In linguistic terms, one thus expects canonical reduplication to involve two syntactic words whose phonological shape is the same quantitatively, qualitatively, and sequentially not only on the segmental but also on the suprasegmental level. Thus, Formula IV is appropriate for segmental phonology whereas Formula V is meant to cover the suprasegmental identity of reduplicand and image.[39]

Formula IV:[40] $[/\pi(x)/_{\text{reduplicand}}\ /\pi(x)/_{\text{image}}]_{\text{canonical reduplication}}$

Formula V:[41] $[/\acute{\alpha}(x)/_{\text{reduplicand}}\ /\acute{\alpha}(x)/_{\text{image}}]_{\text{canonical reduplication}}$

Absolute phonological identity translates into identical morphological structure of the syntactic words. The number, function/meaning, and order of their morphemes are the same for both of the syntactic words. This gives rise to Formula VI.

Formula VI:[42] $[\{\mu\}(\{x\})_{\text{reduplicand}}\ \{\mu\}(\{x\})_{\text{image}}]_{\text{canonical reduplication}}$

On the one hand, prototypical reduplication and canonical reduplication are in agreement as to the criterion of phonological identity. As to morphology, however, no requirement is formulated for the prototype since there is a controversy as to the possibility of images having a morphological structure at all (cf. below). This is another piece of evidence of the dependence of the prototype on the already known empirical facts and the problems they pose for the linguistic analysis. The canonical approach, however, can turn a blind eye to the intricacies of attested cases of reduplication.

If the same morphological analysis can be applied to the syntactic words, they can be considered to be syntactic equivalents of each other too because each of them comes in the appropriate shape that is required by syntax in a

39 I acknowledge that proving prosodic identity of $A_{\text{reduplicand}}$ and A'_{image} is often difficult on the basis of the information provided in the descriptive-linguistic literature on a given language. That is why, throughout Section 4, I desist from giving much prominence to this parameter.

40 The following symbols are employed: π = phonological segment, x = further phonological segments numbering $n \geq 0$.

41 The following symbols are employed: $\acute{\alpha}$ = prosodic property, x = further suprasegementals numbering $n \geq 0$.

42 The following symbols are employed: μ = morph(eme), x = further morph(eme)s numbering $n \geq 0$.

given context. From the point of view of the prototype approach, the supposed syntactic equivalence is questionable since images are often considered to be devoid of syntactic relations of their own (cf. below). For the canonical approach, however, since there is morphological identity and syntactic equivalence of the syntactic words, it hardly needs to be explicated that under canonical reduplication the syntactic words are not only homophonous but also fully synonymous, i.e., they are also identical semantically.[43] This situation is captured by Formula VII.

Formula VII:[44] $\varphi_{\text{reduplicand}} = \varphi_{\text{image}}$

For prototypical reduplication, the semantic equivalence of reduplicand and image has not been promoted to the status of criterion most likely because it is also controversial whether or not images can have content at all (cf. below).

However, this is not the entire story of semantics in the context of prototypical and canonical reduplication. For both concepts, it is a must that the combination of A + A' has a meaning or function which is not that of either A or A' in isolation, cf. Diagram 11.[45]

If the equation $\varphi_{\text{reduplicand}} = \varphi_{\text{image}}$ holds, the most obvious interpretation of $\varphi_{\text{reduplicand}} + \varphi_{\text{image}}$ is that of the increase of the content associated with A and A' separately as shown in $\varphi_{\text{reduplicand}} + \varphi_{\text{image}} = 2\varphi = \varphi_{\text{canonical reduplication}}$. At this point, we enter marshy terrain, in a manner of speaking.

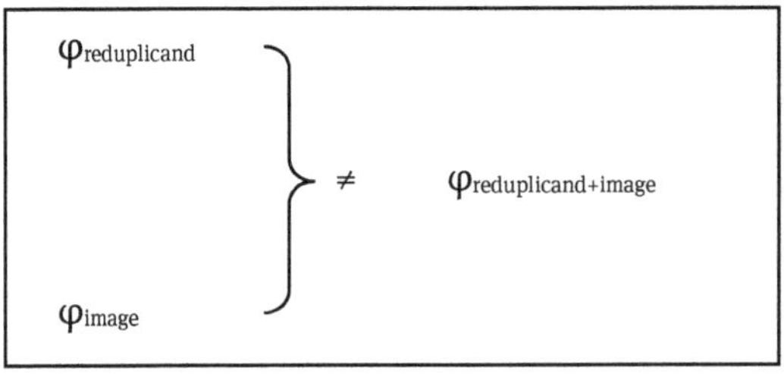

Diagram 11: Functional/semantic difference – singleton vs. reduplication.

---

43 Where Morphological Doubling Theory (MDT) assumes semantic similarity (and not necessarily semantic identity) to be sufficient for a binary construction to pass as an instance of reduplication (Inkelas & Zoll 2005: 7), my approach conceives of canonical reduplication as involving entire linguistic signs (or form-function pairs) so that the two levels of content and expression are required to participate in the process of copying. I argue that it is difficult if not downright impossible to derive a viable prototype or canonical reduplication on the basis of the axioms of MDT because this theory is far too liberal as to what is understood by similarity (Inkelas & Zoll 2005: 47–65).

44 The symbol φ represents functions/meanings.

45 From the point of view of the principle of least effort, it could be asked rhetorically why one should go to the pains of saying something twice when saying it once means the same.

Suppose that $\varphi_{\text{canonical reduplication}}$ simply is the sum of the addition of $\varphi_{\text{reduplicand}}$ + φ_{image}, we get 2φ as result.[46] Finding a way out of this self-imposed dilemma is a demanding task. In this situation, I have recourse to my initial stipulation which assumes construction-status for all instances of canonical reduplication. It follows that canonical reduplication automatically has a meaning/function which is not only different from that of its constituents but also from the sum of the meanings of the constituents. Thus, Diagram 11 can be complemented further to yield Diagram 12.

$$\left.\begin{array}{l}\varphi_{\text{reduplicand}}\\ \\ \\ \varphi_{\text{image}}\end{array}\right\} \neq \quad \varphi_{\text{reduplicand+image}} \quad \neq \quad \varphi_{\text{canonical reduplication}}$$

Diagram 12: Functional/semantic difference – canonical reduplication.

Canonical reduplication yields constructions whose meaning is non-compositional in the sense that they involve at least one semantic feature that is not also there if the meanings of reduplicand and image are added up (Fischer & Stefanowitsch 2006: 5).

Generalizing, it can be said that canonical reduplication reflects iconicity of quantitity in the sense that more of form goes hand in hand with more of content. This is an aspect that has not been deemed to be essential for prototypical reduplication. It is absent from the list of criteria of prototypicality exactly for the same reasons as in the previous cases. We already know that many voices have been raised against taking iconicity as a general characteristic of reduplication because

46 If this were really true, the dual number would be a prime candidate for being expressed by means of reduplication. This expectation is not corroborated by the empirical facts. Dual number is only very rarely expressed by reduplication – as Fabricius (1998: 71) complains. One of the infrequent examples I am aware of is given by Rubino (2005a: 115 = 2005b: 20) who mentions Luiseño (Uto-Aztecan) *lawi* 'to make holes' → DUAL *law-lawi* 'to make two holes; to make a hole twice' ≠ PLURAL *lawa-láwi* 'to make more than two holes'. I repeat, however, that empiry should not have an impact on canonicity, meaning: the scarcity of the dual being expressed by reduplication does not principally preclude the possibility to declare this number category the canonical function of reduplication. The reason why I do not opt for the dual is of a different nature. The meaning of the construction is (strictly) compositional if the combination of reduplicand and image results in a dual reading.

there is empirical evidence of reduplication being used non-iconically or even counter-iconically (Kouwenberg & LaCharité 2015, Rozhanskiy 2015). The possibility of an alternative conception of iconicity notwithstanding (Stolz 2007b),[47] it is clear that canonical reduplication is exempt from taking any precautionary measures of the kind prototypical reduplication is forced to take on account of extant theories and problematic empirical evidence.

In sum, canonical reduplication has the following properties:[48]

(C1.) $A_{reduplicand}$ is copied as A'_{image};
(C2.) $A_{reduplicand}$ precedes A'_{image};
(C3.) $A_{reduplicand}$ and A'_{image} are immediately adjacent to each other;
(C4.) multiple images are ruled out;
(C5.) $A_{reduplicand}$ and A'_{image} are syntactic words;
(C6.) $A_{reduplicand}$ and A'_{image} are identical prosodically
(C7.) $A_{reduplicand}$ and A'_{image} are identical phonologically (segmental level);
(C8.) $A_{reduplicand}$ and A'_{image} are identical morphologically;
(C9.) $A_{reduplicand}$ and A'_{image} are identical syntactically;
(C10.) $A_{reduplicand}$ and A'_{image} are identical semantically;
(C11.) the meaning/function of the reduplicative construction is different from both that of $A_{reduplicand}$ and that of A'_{image};
(C12.) the meaning/function of the reduplicative construction is different from that of the sum of meanings/functions of $A_{reduplicand} + A'_{image}$;
(C13.) the meaning/function of the reduplicative construction is associated with the iconicity of quantity.

The requirements of prototypical reduplication and canonical reduplication are the same for several of the criteria (C1–C13). The absence of an equivalent of criterion (vi) of the prototype in the list of canonical properties is directly derivable from criterion (C5) of canonical reduplication because syntactic words per definition do not have a host so that they cannot cause discontinuity of other word units. Arguably, the criteria (C6–C9) could be subsumed alternatively under the umbrella of a parameter which states that $A_{reduplicand}$ and A'_{image} are identical in all aspects of the expression side.

47 There is of course also Haspelmath's (2008) general criticism of the uses the notion of iconicity has been put to in linguistics. Since in studies on reduplication, however, this incriminated notion is still much en vogue, I feel justified to refer to it in spite of its many deficiencies.

48 To distinguish these properties from those of prototypical reduplication exposed in Section 3.1, I use Arabic numerals in the case of canonical reduplication.

As to the parameters (C8–C10), some clarification is required. In the previous paragraphs I have repeatedly mentioned that images are often denied internal morphological structure, a syntactic status and a meaning of their own.[49] This is why images are frequently glossed as RED (or the like) without acknowledging that they might be morphologically complex and thus justify being segmented further.[50] This practice can certainly be defended in the vast majority of phenomena which go by the name of partial reduplication. For canonical reduplication, however, the exigencies of partial reduplication are irrelevant because I envisage a construction that consists of two full-blown syntactic words so that reduplicand and image are in a relation of complete and exact identity – a relation of full equality which implies that both constituents of the construction are invested with the entire set of properties as required by the category they represent. As far as morphology, syntax, and semantics go, my canon has very much in common with the concept of identical daughters as propagated in the framework of MDT (Inkelas & Zoll 2005: 11–15). This, however, is already the extent of the resemblance of the two approaches. For the requirements of canonicity to be met, it is necessary that also prosodical and segmental identity applies whereas the proponents of MDT explicitly challenge the necessity of phonological likeness for reduplication (Wälchli 2007: 102–103). Furthermore, in line with their generativist convictions, Inkelas & Zoll (2005: 7) postulate a

49 In contrast, Dingemanse (2015) glosses the image as a unit which bears a meaning of its own as in (ii) from Siwu.

(ii) Siwu [Dingemanse 2015: 948]

kàde	*kawɛ̃*	*nɛ,*	*ma-ɛɛ*	***màfuri~mafùrì***
land	certain	TP	ma-HES	**PL.albino~DISTRIB**

'There is a land, a-uhm **albinos here and there**'

The reduplicand *màfuri* 'albinos' is a syntactic word in the plural. Its image is prosodically different from the reduplicand but segmentally identical with it. Going by the morpheme glosses provided by my source the image itself conveys the meaning of the distributive. In my approach, it is the entire reduplicative construction which expresses the distributive.

50 Occasionally Fabricius (1998) glosses morphologically complex images as REDUP (= reduplication) – probably according to the practice of the original sources from which she draws the examples. A case in point is (iii).

(iii) Kriol [Fabricius 1998: 51]

olabat	*bin*	***graj-im~grajim***	*yem*
3PL	TNS	**dig-TRSV~REDUP**	yam

'They were **digging** yams'

The reduplicand *grajim* 'to dig' is a syntactic word which hosts the transitivizer *-im*. This is dutifully marked in the above morpheme glosses. Its phonologically identical image *grajim*, however, is presented as an unanalyzable chain of segments without meaning although taken in isolation it would call for the very same morphological segmentation as the reduplicand.

generally underlying pattern of morpho-semantically identical constituents of a reduplicative construction on the basis of which the surface realizations can be generated by way of applying (especially though not exclusively) phonological rules which affect the underlying constituents to varying degrees to create the correct output.[51] Canonicity does in no way presuppose an underlying pattern since it purposefully leaves open the question of whether or not the phenomena under review are derivable from a common source – be it an underlying structure or – especially in the light of the skepticism expressed by Hurch & Mattes (2005: 153–154) – a grammaticalization path. The interconnection of the reduplicative constructions and their look-alikes can be better argued for or against after we have checked what order arises from the application of the principles of the canonical approach.

I do not consider this list of a dozen properties to exhaust the characteristics of canonical reduplication. For the time being, however, I make do with this first attempt at determining the canonicity in the domain of reduplication because it provides sufficiently robust foundations to contrast canonical and non-canonical reduplication in Section 4.

4 Canonicity put to practice

In this section, I look more closely at empirical data to determine in what way and to what extent attested instances of reduplication may deviate from the ideal of canonical reduplication and thus testify to non-canonicity. In a like-minded manner, Mattes (2014: 35–40) sketches an embryonic version of what in the future might develop into a full-blown matrix of criteria for the classification of phenomena in the domain of reduplication. I do not intend to provide a comprehensive catalogue of what the languages of the world have on offer in the realm of reduplication. A recent cross-linguistic survey of the phenomenology of total reduplication can be found in Kallergi (2015: 390–404); partial reduplication is amply documented in the Graz Database on Reduplication (http://ling.uni-graz.at/reduplication/) (Hurch & Mattes 2009).

---

51 Note also that the Formulas IV–V are largely in line with the Full-Copy Theory (on which Inkelas & Zoll 2005: 68 elaborate). As with the axioms of MDT generally, there is the problem that Full-Copy Theory postulates an underlying structure from which all other reduplicative constructions can be generated. This postulate is alien to the canonical approach.

In Section 4.1, I start from an evaluation of total reduplication since it is often depicted as a hot candidate for the status of a universal of human language – although the issue is a particularly controversial one (Stolz et al. 2015: 797–798). The would-be universality makes total reduplication a suitable comparee of canonical reduplication (Otsuka 2016: 21). Section 4.2 is dedicated to the cursory discussion of echo-word formation as one of the attested deviations from canonicity. On the basis of the findings in Section 4.1–4.2, I delineate the boundaries of canonical reduplication in relation to non-canonical reduplication.

4.1 Total reduplication

The above list of criteria (C1–C13) is largely reminiscent of what is usually termed (syntactic) total reduplication. Superficially, the prototype of reduplication seems to invoke the notion of total reduplication too. However, canonical reduplication gives preference to a variety of total reduplication which is only one of several options in the case of prototypical reduplication.

4.1.1 Types of total reduplication

According to Rubino (2005a: 11, 2005b: 114)[52], total reduplication (= his full reduplication) may come in three shapes, namely as the "repetition" of

- an entire word,
- a word stem (root with one or more affixes),
- a root.

What the three options have in common is that they always involve the entire lexical morpheme of a given word unit. Reduplication thus is total if and only if that part of a word which conveys lexical meaning is fully represented in $A_{\text{reduplicand}}$ and A'_{image} alike. A convenient representation is Formula VIII.

Formula VIII:[53] $[\{\text{LEX}_{\text{reduplicand}}\}(+x){\sim}\{\text{LEX}_{\text{image}}\}(+y)]_{\text{total reduplication}}$

52 The two papers of Rubino's (2005a–b) largely overlap as to the descriptive text they contain. Their close correspondence in terms of wording notwithstanding, I reference both papers side by side because the empirical examples used to illustrate reduplicative phenomena are not always the same so that making use of both sources help to get a better grip on what the author has in mind when he talks about certain facts.

53 The following conventions hold for this formula: LEX = LEX; x and y are variables for any numbers of optional additional (non-lexical) morphological units which are not necessarily

The compulsory lexical nature of part of the input of the reduplicative process is not an explicit requirement of prototypical reduplication and canonical reduplication. Since both the prototype and the canon assume the status of syntactic word for $A_{reduplicand}$, it may be said that lexicality is at least implicitly required too if syntactic words are supposed to always involve a lexical morpheme as well.

The lexicality condition is probably too restrictive. Mattes (2014: 35) adds affix reduplication to the above list of types of total reduplication with the proviso that the reduplication of affixes is attested infrequently across languages. To her mind "[f]ull reduplication means that a morphological constituent (the simplex form) is copied as a whole" (Mattes 2014: 35) without restricting this process to morphological units which carry lexical meaning. Since this type of reduplication is absent from Bikol – the language studied by Mattes (2014) – the author does not provide examples of affixes which undergo total reduplication. Examples can be found, however, in Van der Voort's (2014: 448–449) account of "morphologically-based reduplication of bound morphemes" in the Amazonian isolate Kwaza. The process involves bound person markers which undergo reduplication "to indicate remote past or habitual aspect". As the Kwaza example (9) shows the disyllabic bound person marker *-axa* of the 1st person plural exclusive is copied in its entirety.

(9) Kwaza [Van der Voort 2014: 449]
*aure-lɛ-nã-**axa~'axa**-le-hĩ-ki*
marry-RECI-FUT-**1EXC~1EXC**-FRUST-NOM-DEC
'**We were** going to marry (but we didn't)'

The phonological chains of reduplicand and image differ because the latter has an initial glottal stop (= <'>) which is absent from the reduplicand. However, this is not a case of fixed segmentism but the automatic insertion of the consonant to avoid sequences of vowels. Therefore, reduplicand and image are phonologically identical.

Van der Voort (2014: 448–449) is adamant that Kwaza affix reduplication is by no means phonologically induced. In Diagram 13, I contrast the morphological structure of the syntactic word (as representative of the canonical input) with the image to show that the Kwaza case is far removed from canonicity because otherwise all the empty cells of the image would be filled identically to the corresponding cells of the syntactic word.

identical for $A_{reduplicand}$ and A'_{image}. For simplicity, I employ a formula which is suggestive of suffixation although, mutatis mutandis, the principles on which it is based hold for all kinds of morphologies.

Syntactic word	*aure*	*lɛ*	***axa***	*le*	*hĩ*	*ki*
Image			***axa***			

Diagram 13: Kwaza affix reduplication.

Another case in point is Hungarian (Uralic) preverb reduplication (Inkelas & Zoll 2005: 28) which cannot be explained with reference to phonology either. The Hungarian preverbs are variously classified as prefixes or proclitics (Forgács 2007: 244–245). Since the preverbs are mobile and are often separated from the verb they cannot be considered to be bound morphemes. The examples in (10) illustrate the reduplication of the preverb *ki-* as well as the separability of the same preverb from the lexical verb (= *menni* 'to go' in (10b)).

(10) Hungarian (Forgács 2007: 247)[54]

a. ***Ki~ki**-néz* *az* *ablakon*
out_of~out_of-look DET window:SUPERESS
'**Again and again** he looks out the window'

b. *Ki* *akarsz* *menni* *a* *szobából?*
out_of want:2SG go:INF DET room:ELA
'Do you want to leave the room?'

The reduplication of preverbs serves the purpose of expressing the frequentative-iterative. Preverbs may be monosyllabic like *ki-* or disyllabic like *vissza-* which translates to English *back*. The number of syllables is no obstacle to reduplication. I have no evidence of the reduplication of preverbs when they are separated from their verbal host.

In Diagram 14, I contrast the segmental chain of the syntactic word with that of the image. The extra-bold line marks the morpheme boundary.

Syntactic word	***k***	***i***	*n*	*é*	*z*
Image	***k***	***i***			

Diagram 14: Hungarian preverb reduplication.

54 The morpheme glosses and the English translation are mine.

As in the above case from Kwaza, the distance from the assumed canon is evident. Clitics form a category which is neither fully lexical nor affixal in nature and thus needs to be distinguished from words on the one side and properly bound morphemes on the other (Haspelmath & Sims 2010: 196–203). Like affixes, roots, stems, and words, clitics may be subject to reduplication. Since they are morphological constituents of (phonological) words, they are thus candidates for a fifth type of total reduplication. Rubino's above ternary set of types of total reduplication gains in size by way of adding two more types to the list,[55] namely total reduplication of

- an affix, and
- a clitic.

This extension of the types of total reduplication calls for Formula IX which is meant to cover affix and clitic reduplication.[56]

Formula IX:[57] $[\{\text{LEX}\}\text{-}\{\text{AFFIX/CLITIC}_{\text{reduplicand}}\}{\sim}\{\text{AFFIX}_{\text{image}}\}]_{\text{total reduplication}}$

55 In the light of the findings of Nintemann (2016, this volume) and Robbers (2016) it makes sense to assume that the traditional strict bipartition of the domain of reduplication in total vs. partial reduplication (Mattes 2014: 35) is oversimplifyingly inadequate because it glosses over phenomena whose particulars need to be inquired into to come to grips with the rich variety of shapes in which reduplication is attested. The above mentioned studies show that constraints which determine the size of the image in terms of the maximum number of syllables allowed are widely common cross-linguistically. These constraints yield a type of reduplication that can be classified straightforwardly neither as total reduplication nor as partial reduplication. It is tempting to speculate that the concept of canonical reduplication raises further issues that are suggestive of the necessity to revise the traditional categorization. This, however, is a topic that deserves to be addressed in a separate study in the future.

56 The integration of affix and clitic reduplication in the list of types of total reduplication yields far-reaching implications. It is common to assume that stem and root reduplication take place prior to inflection as shown in for instance Da Cruz's (2014: 120–121) study of reduplication of Nheengatu (Tupi). The input *yuká* 'kill' is first totally reduplicated as *yuka~yuka* before the agent prefix *u-* of the 3rd person singular is added to yield the surface form *u-yuka-yuka* 'He kills repeatedly'. However, for proper affixes to function as reduplicand the word-form must be inflectionally complete, i.e. the reduplication takes place post-inflectionally. Given this analysis, it remains to be determined whether or not the reduplicand (=affix) and the domain (=affix or syntactic word) are one. If they are not, the distinction of total and partial reduplication is again at stake.

57 The following conventions hold for this formula: AFFIX = AFFIX; CLITIC = CLITIC. As in previous formulas, I represent the internal structure of the construction according to the patterns of a suffixing language.

In the light of the concept of canonicity, I interpret this list of five types of total reduplication as a scale of decreasing canonicity in lieu of treating them as egalitarian competitors. Total reduplication of entire words is fully in line with canonical reduplication because the above Formula III (= $[\omega_{\text{reduplicand}}\ \omega'_{\text{image}}]_{\text{canonical reduplication}}$) captures this variety of total reduplication nicely. The other four options, however, are less easily accommodated to the concept of canonical reduplication. The major reason for this problem is of course the limited scope the further varieties of total reduplication have over the syntactic word. Total reduplication of stems, roots, affixes, and clitics fails to cover the entire chain of segments which constitute a syntactic word.

If stems are reduplicated totally there may be some surplus morphology that is exempt from being copied so that the following Formula X provides an adequate schematic rendering of the output.

Formula X:[58] $[\|\{\text{LEX}\}\text{-}\{\mu_{\text{deriv}}\}(\text{-x})|_{\text{reduplicand}}\sim\{\text{LEX}\}\text{-}\{\mu_{\text{deriv}}\}(\text{-x})|_{\text{image}}\text{-y}\|_{\omega}]_{\text{total reduplication}}$.

In the case of total reduplication of roots the appropriate Formula XI looks different.

Formula XI:[59] $[\|\{\text{LEX}\}_{\text{reduplicand}}\sim\{\text{LEX}\}_{\text{image}}(\text{-x})|\text{-y}\|_{\omega}]_{\text{total reduplication}}$.

If we discount the implicit criterion of lexicality and look at Formulas IX–XI from the point of view of the syntactic word as the basic meaning bearing unit which is affected by reduplication, what is copied covers only a part of the phonological chain of the entire syntactic word.

We note that, when compared to canonical reduplication, the five supposed types of total reduplication are too different from each other to be treated as one. On the continuum as well as in the radial category of reduplication, the different options occupy different positions relative to canonical reduplication which either constitutes one of the extreme poles of the continuum or the center of the radial category. Incidentally (or not), the order in which the first three options are introduced in Rubino (2005a–b) reflects this scale very closely. The total reduplication of affixes and clitics has been illustrated in (9)–(10) above. Therefore, Section 4.1.2 focuses on the total reduplication of words, stems, and roots.

---

58 For expository reasons, my Formula X reflects exclusively the patterns of suffixation. In this formula I assume that LEX = LEX and μ = μ holds. Lower case x represents any number of further derivational morphemes whereas y indicates that there is inflectional morphology only in the rightmost slot. The symbol ‖ marks the boundaries of the syntactic word whereas | designates the right margin of stems.

59 The conventions are the same as those used for the previous Formula X.

4.1.2 Nweh, Basque, and Quechua

On the topmost rank, there is the reduplication of entire words. Rubino characterizes his examples of total reduplication of entire words as lexicalized cases, namely

- Tausug (Austronesian) *dayang* 'madam' → *dayang~dayang*[60] 'princess' (Rubino 2005a: 11),
- Tausug (Austronesian) *laway* 'saliva' → *laway~laway* 'land snail' (Rubino 2005a: 11),
- Nez Perce (Sahaptian) *té:mul* 'hail' → *té:mul~té:mul* 'sleet' (Rubino 2005b: 114).

The so-called entire words which provide the input do not host any bound morphology. The absence of examples of totally reduplicated inflected entire words strikes the eye especially because the second option of total reduplication is declared to involve roots plus an unspecified number of affixes (though never the entire array of bound morphology of a given word-form). In this context, it is remarkable that the second variety of total reduplication (= reduplicated stems) is not illustrated by empirical proof in Rubino's papers to which I refer. If the reduplication of a root along with (some of) its affixes is possible, one might expect that also morphologically complex i.e. inflected "entire" words can form the input of total reduplication.

This assumption is corroborated by the empirical facts. Consider example (11) from Nweh (Grasslands Bantu).

(11) Nweh [Njika 2012: 111][61]

a. *à* *nǝ̌ŋ* *bə̀-kunyà*
s/he rear PL-pig
'S/he rears pigs'

b. *à* *nǝ̌ŋ* ***bə̀-kunyà~bə̀-kunyà***
s/he rear **PL-pig~PL-pig**
'S/he rears **exclusively pigs** (and nothing else)'

The noun *kunyà* 'pig' is overtly marked for plural number in (11a–b). In (11b), the number/class prefix *bə̀-* is used not only with the reduplicand but also with

60 In this and the two following examples, I have added the sign ~ to indicate the boundary between reduplicand and copy.

61 In contrast to my source, I have glossed the subject pronoun in a gender-neutral way. Moreover, I have marked the morpheme boundary for the noun in (11a). In (11b), I have replaced the original hyphen with the sign ~ to join reduplicand and copy.

the image so that it appears twice in the construction. Diagram 15 shows that all segments of the syntactic word (= reduplicand) have an exact replica in the image. The morphological structure of reduplicand and image is the same, too.

$A_{reduplicand}$	*b*	*ə̀*	*k*	*u*	*ny*	*à*
A'_{image}	*b*	*ə̀*	*k*	*u*	*ny*	*à*

Diagram 15: Total reduplication in Nweh.

The degree of canonicity of the Nweh example is indisputably high. Njika (2012: 111) herself draws the reader's attention to the fact that

> [u]nlike the non[-]inclusion of the tense marker in the verbal reduplicant, the plural marker is an integral part of nominal reduplicants.

Morphologically marked plurality is obligatory for totally reduplicated nouns if the intended meaning is that of "unspecified plurality" or of "large quantity of specific objects (both animate and inanimate) to the absolute exclusion of others" (Njika 2012: 110).[62]

According to the same author, in (11b), $A_{reduplicand}$ and A'_{image} form a complex word-unit – at least if we go by the orthographic practice of joining the two constituents by a hyphen in the source. If we take orthography at face value, this means that Njika (2012) does not classify the reduplicative construction under scrutiny as a syntagm. What we are dealing with is thus the total reduplication of an inflected syntactic word yielding a doubly inflected compound-like syntactic word. This pattern is systematically employed and has a constructional meaning. Njika (2012: 112) states that

> [a]s in the reduplication of verbs, Nweh nouns[sic!] reduplication is also restricted to duplicating the reduplicant once; and it yields both semantic and grammatical meaning.

The Nweh case is thus not far removed from canonical reduplication.

I assume that examples like (11b) are absent from Rubino's (2005a–b) surveys mainly because there is bound affixal morphology exactly on the boundary between reduplicand and image. The presence of the number/class prefix on

62 In other reduplicative constructions of Nweh, nouns remain in the singular as, e.g., when they serve as the basis of adverbial or adjectival derivation: *àkpàŋ* 'a joke' → *àkpàŋ-àkpàŋ* 'jokishly' and *àmʉ̀'* 'mist' → *àmʉ̀'-àmʉ̀'* 'misty' (Njika 2012: 112).

the right constituent of the construction (which most probably is the image[63]) is problematic for an approach that restricts the domain of reduplication to the word-level. On the one hand, word-internal inflectional morphology is a very marked phenomenon; inflections are normally expected to occupy slots on the margins of a given word-form (Haspelmath 1993). Thus, the number/class prefix invokes a word boundary. On the other hand, treating $A_{reduplicand}$ and A'_{image} as a construction which comprises two syntactic words means that they form a syntagm – and this in turn means that constructions of this kind fall outside the scope of the word-based approach. There is thus a dilemma.

The same dilemma arises in the case of Basque. De Rijk (2008: 881) makes a terminological distinction between reduplication and duplication – both of which are considered iterative phenomena:

> This duplication, only applied to place and time indicators with a locational ending, differs in form and meaning from the reduplication discussed previously – in form because the ending is also repeated, and in meaning because the repetition does not intensify or strengthen the base form. Instead, it gives a distributive, generalizing dimension to the word.

As shown in (12), case suffixes behave differently under the two kinds of reduplication.

(12) Basque[64]

a. "duplication" [De Rijk 2008: 882]

Urte-a-n	***urte-a-n***	*igarotzen*	*dute*	*hementxe*	*uda*
year-DEF-LOC	**year-DEF-LOC**	pass:ITERV	AUX.3PL.ERG	here:EMPH	summer

'**Year in year out** they spend the summer here'

b. "reduplication" [De Rijk 2008: 878]

Lo	***gozo~gozo**-a-n*	*nengoen*
sleep	**sweet~sweet**-DEF-LOC	1SG.PRET:be

'I had fallen into the **most blissful** sleep'

Especially as to the formal aspects of reduplication, example (12a) is reminiscent of example (2) from Oriya. Two identical inflected syntactic words combine to yield a construction with distributive meaning. The reduplicand as well as the

63 This assumption is based on the observation that my source contains a number of examples of derivationally employed reduplication of nouns in which the right constituent has prosodic properties other than those of the left constituent which preserves the suprasegmental structure of the singleton item as, e.g., in *lènyĭn* 'headiness' → *lènyĭn-lenyin* 'in a heady manner' (Njika 2012: 112).

64 The morpheme glosses are mine.

image host the locative suffix and, in the Basque case, both $A_{reduplicand}$ and A'_{image} are overtly marked for definiteness. Since Basque also attests to reduplication with segmental variation on the image in the right slot[65], I assume that the linearization of $A_{reduplicand}$ and A'_{image} applies and thus is canonical. Like its Oriya counterpart, the Basque example in (12a) thus rates relatively high in terms of canonicity. This can be gathered additionally from the schematic rendering in Diagram 16 which has identically filled cells for reduplicand and image.

$A_{reduplicand}$	*u r t e*	*a*	*n*
A'_{image}	*u r t e*	*a*	*n*

Diagram 16: Total reduplication of inflected words in Basque.

In contrast, the reduplicative construction in (12b) involves two identical chains of segments which are inflected only once, namely on the right margin where we find the definiteness marker -a together with the locative suffix *-n*. According to Rubino's threesome of types of total reduplication, (12b) can be identified as an instance of root reduplication according to the following succession of steps:[66]

- input: *gozo*$_{root}$ 'sweet'
  - reduplication of root:
    - *gozo*$_{reduplicand}$~*gozo*$_{image}$ 'very sweet'
  - inflection:
    - [[{*gozo~gozo*}]$_{total\ reduplication}$-{*a*}-{*n*}]$_{\omega}$ 'in the very sweet'

65 I refer to cases like *duda* 'doubt' → *duda*$_{reduplicand}$~*muda*$_{image}$ 'all sorts of qualms' (with obligatory intitial /m/ on the copy) called expressive compounds by De Rijk (2008: 864–867).

66 Note, however that De Rijk (2008: 880) states that "[n]ouns denoting a time, place, or situation can be reduplicated, but only when they are accompanied by an appropriate case ending." This statement invokes the order of inflection preceding reduplication. The same author refers to cases like the postpositional phrase *etxe aurrean* 'in front of the house' → *etxe* ***aurre-aurrean*** **'right in front of** the house' which involve the nouny postposition *aurre* 'in front of' that undergoes reduplication in its fully inflected form (*aurrean* = definite locative 'at the front of'). It seems to be plausible that the original construction from which the reduplicated output is produced is the postpositional phrase *etxe aurrean* 'in front of the house' and not **etxe aurre-aurre.*

The additional string of segments *gozo-* can be shown to be phonologically, morphologically, semantically, and syntactically different from the syntactic word gozoan 'in the sweet' (cf. Diagram 17).

Syntactic word	*g*	*o*	*z*	*o*	*a*	*n*
Image	*g*	*o*	*z*	*o*		

Diagram 17: Root reduplication in Basque.

The constellation of facts of Diagram 17 is unproblematic for an approach which assumes the above order of steps, i.e. reduplication precedes inflection. In the context of canonicity, the situation changes in the sense that the output does not consist of two syntactic words. This means that total reduplication of roots fails to fulfill a basic requirement of canonicity so that, in contrast to the example of total reduplication of inflected words in (12a), (12b) is clearly more remote from the canonical type.

Evidence for the total reduplication of stems can be found in Tarma Quechua.[67] Example (13) involves a stem that consists of the root *šarku* 'to stand up' and the agentivizer suffix *-q*. The stem *šarkuq* 'one who stands up' is already a possible syntactic word.

(13) Tarma Quechua [Hannß & Muysken 2014: 62]
šarku-q ***šarku-q****-la=m* *ka-ya-n*
stand_up-AG **stand_up-AG**-LIM=DECL this-ATT-3
'He wants **to get up all the time**'

67 For Basque, De Rijk (2008: 880) specifically states that with a morphologically complex $A_{reduplicand}$ "[t]he locative ending *-tan* does not reduplicate, but the word-formation suffix *-ero* does." The preservation of the derivational affix *-ero* under reduplication does not yield an output that can be taken to illustrate Rubino's total reduplication of stems because no further inflections can be added. In (iv), the reduplicand as well as the image host this affix (the morpheme glosses and the English translation are mine).
(iv) Basque [Bendel 2006: 193]
Ast-ero~ast-ero *igerilekura* *joaten* *naiz*
week-ADV~week-ADV swimming_pool:ALL go:PROG AUX.1SG.ABS
'I go to the swimming-pool **every week (without exception)**'
The adverb *astero* 'every week' is derived from the noun *asti* 'week'. The derivative *astero* does not only convey an iterative meaning but it is also a full-blown syntactic word itself.

This stem occurs twice in the example with only the rightmost constituent hosting the affixes of the limitative and the declarative. If we take *šarkuqlam* to be the syntactic word against which the canonicity of the reduplication can be measured, it is clear that the total reduplication of the stem alone falls short of meeting the expectations we have of instances of canonical reduplication. The discrepancy results clearly from Diagram 18.

Syntactic word	*š*	*a*	*r*	*k*	*u*	*q*	*l*	*a*	*m*
Image	*š*	*a*	*r*	*k*	*u*	*q*			

Diagram 18: Stem reduplication in Tarma Quechua.

In a variety of languages, roots may or must reduplicate alongside further morphological units. For her evidence from Australian languages, Fabricius (1998: 49) assumes that "[t]hese examples seem to arise due to a basic 'syllabicity condition' applying in the language." Similar cases are reported from East Africa by Nintemann (2016: 15–20, this volume).[68]

The discussion of the above examples indicates that the different types of total reduplication are associated with different degrees of canonicity. In Table 3, I tick off each of the parameters of canonicity as presented in Section 3.3 above. The judgments refer exclusively to the examples (9)–(13) discussed above. The symbol √ is used if a criterion is met. The same symbol appears in brackets if the analysis leaves a margin of doubt. I employ **no** if a criterion is violated against. The bracketed negation (= **(no)**) signals that the criterion is (perhaps) not applicable in the first place. The results are provisional and should therefore be taken with a grain of salt.

Table 3: Check of canonicity with types of total reduplication.

Criterion	Basque (12a)	Nweh (11b)	Quechua (13)	Kwaza (9)	Hungarian (10a)	Basque (12b)
	√	√	√	√	√	√
	√	√	√	√	√	√
	√	√	√	√	√	√
	√	√	√	√	√	√

68 Cf. Robbers (2016: 194) for comparable cases in the Americas.

Criterion	Basque (12a)	Nweh (11b)	Quechua (13)	Kwaza (9)	Hungarian (10a)	Basque (12b)
	√	(√)	√	no	no	no
	√	√	√	no	no	no
	√	√	(√)	(no)	(no)	(no)
	√	√	(√)	(no)	(no)	(no)
	√	√	(no)	(no)	(no)	(no)
	√	√	(no)	(no)	(no)	(no)
	√	√	√	√	√	√
	√	√	√	√	√	√
	√	√	√	√	√	√

On the basis of Table 3, it is possible to picture the situation in the format of a radial category. Diagram 19 helps us to better visualize the differential canonicity of the examples (9)–(13). The numbers in Diagram 19 refer back to the above examples.

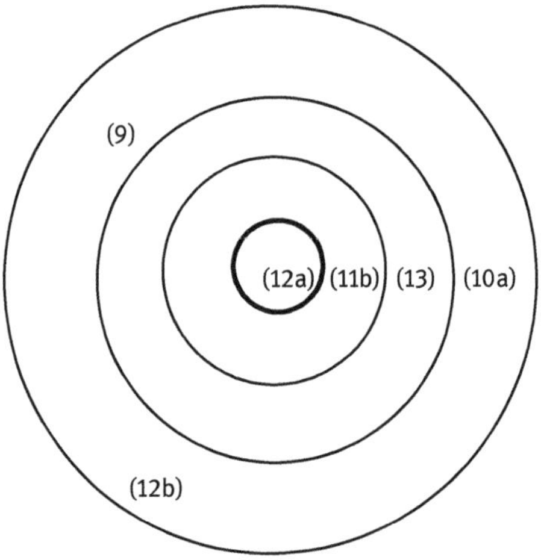

Diagram 19: Radial category of total reduplication.

In terms of canonicity, not everything that passes as total reduplication in the literature fares particularly well. The examples (12a) from Basque, (11b) from Nweh, and (13) from Quechua are privileged in the sense that they fulfill most of the requirements of the canon whereas all other examples display much less impressive scores. Between Quechua, on the one hand, and Kwaza, Hungarian, and example (12b) from Basque, on the other hand, there is a sizable gap so that it can be concluded that the reduplication of affixes, clitics, and roots is not a very good representative of canonical reduplication. Stem reduplication seems to be less of a problem provided that the stem may also have full word status.

The different degrees of canonicity shown in Diagram 19 guide us directly to the subsequent Section 4.2 in which I address the issue of measuring non-canonicity.

4.2 Canonicity vs. non-canonicity

In the domain of morphology, the canonical approach operates with the notion of (form-function) mismatch (Corbett 2005: 34). Mismatches manifest themselves in the failure of attested constructions to fulfill the requirements of canonicity. A single deviation from the canonical type is sufficient to reduce the degree of canonicity and at the same time to increase the degree of non-canonicity of a given linguistic object. I am not entirely convinced that it is appropriate to employ the term mismatch also in the case of non-canonical instances of reduplication. For the time being, I therefore make use exclusively of the term non-canonical and its derivations. An array of candidates for the status of non-canonical reduplication has been the topic of Stolz (2007a: 54–77) and Stolz et al. (2011: 57–68) with some of the discussed phenomena winding up in the waste-paper basket category of false friends. In this paper, however, I do not intend to discard the non-canonical cases sweepingly. The aim of the subsequent sections is to highlight the yard-stick function of the canon when it is applied in the process of evaluating non-canonicity. To this end, I take my pick from the rich empiricism of (putative) reduplicative constructions which do not fully correspond to canonical reduplication. It is not my intention to provide a comprehensive catalogue of non-canonicity in the realm of reduplication. The compilation of a catalogue of this kind is an almost Sisyphean labor which cannot be tackled here and now. Therefore, the focus of Section 4.2.1 is on echo-word constructions. Section 4.2.1.1 highlights some aspects of simple cases of echo-word formation whereas Section 4.2.1.2 takes a look at echo-word formation patterns which display a cumulation of non-canonical properties. To conclude the empirical survey, Section 4.3 touches upon a selection of further issues, namely partial reduplication (= Section 4.3.1) and ultimately sundry phenomena which usually contribute to making the study of reduplication difficult if it is meant to be all-embracing.

4.2.1 Echo-word formation

Echo-word constructions can be encountered in a sizable number of languages as shown in Southern's (2005: 239–244) study of the supposed spread of this construction type from Turkic to other languages in a chain-reaction of language contacts. According to Kallergi (2015: 18) this kind of construction can be described as

> a case of word reduplication involving both morphonological alternation and a pseudo-morpheme. It in fact consists in the copying of a word and the addition to or substitution of the first consonant, consonant cluster or syllable of that copy with another consonant or fixed segment.

The presence of echo-word formation patterns throughout South Asia is a widely acknowledged fact (Southern 2005: 135–156). Interestingly, Abbi (1992: 20) considers echo-word formations to constitute cases of partial reduplication since part of the reduplicand's chain of segments is replaced with phonological material that is proper to the image. Stolz et al. (2011: 47) consider echo-words to represent total reduplication. On closer inspection, the classification of echo-word formations as partial reduplication is not as straightforward as it seems.

4.2.1.1 From simple to more complex cases

In the literature, the paradigm case of a language that attests echo-word formation is Turkish. In (14) I provide two examples which are representative of the bulk of the evidence from modern Turkish.

(14) Turkish[69]

a. initial consonant replacement [Müller 2004: 53][70]

dergi	***<u>m</u>ergi***	*okumuyor*
newspaper	**<u>m</u>:ECHO**	read:NEG:PRES

'He does not read **newspapers and such like**'

b. prothetic consonant [Müller 2004: 55]

bahçede	***ağaç***	***<u>m</u>ağaç***	*yok*
garden:LOC	**tree**	**<u>m</u>:ECHO**	NEG.EXI

'In the garden, there is no **tree or anything of that kind**'

69 The morpheme glosses and English translations are mine.

70 This example is originally from Lewis (1969: 237).

The singly underlined fixed segment /m/ is an exclusive property of the image. As to the echo-word construction in (14a), one might argue that reduplicand and image yield the same number of segments so that there is phonological reduplicand-image correspondence at least quantitatively. However, example (14b) involves an echo-word construction the image of which contains a consonant in addition to the entire chain of segments of the reduplicand. The differences between the two echo-word constructions are clearly visible from a confrontation of Diagrams 20 and 21. The obligatory initial bilabial nasal /m/ of the image is marked out in boldface.

$A_{reduplicand}$	*d*	*e*	*r*	*g*	*i*
A'_{image}	**m**	*e*	*r*	*g*	*i*

Diagram 20: Constituents of Turkish echo-word constructions I.

$A_{reduplicand}$		*a*	*ğ*	*a*	*ç*
A'_{image}	**m**	*a*	*ğ*	*a*	*ç*

Diagram 21: Constituents of Turkish echo-word constructions II.

In Diagram 20, the initial /m/ of the image corresponds to the voiced dental plosive /d/ of the reduplicand whereas in Diagram 21, the onset of the initial syllable of the reduplicand is empty, i.e., the word has an initial vowel. Since the slot of the onset of the image is filled by the compulsory /m/, the chain of segments of the image exceeds the length of the chain of segments of the reduplicand by exactly one segment. In the absence of consonantal replacements, the entire segmental chain of the reduplicand is copied to form part of that of the image. One might ask whether, in the latter case, it is legitimate to speak of total reduplication. A positive answer to this question creates a grave problem, namely that of dissociating the two patterns of echo-word formation from each other because one is treated as an example of partial reduplication whereas the other counts as total reduplication. This paradoxical situation calls into question the legitimacy of the categorically binary distinction of partial and total reduplication.

In the Turkish case, the rule which covers the vast majority of the echo-word formations consists in creating an obligatory consonantal onset /m/ for the initial syllable of the image. Words which display an initial /m/ already in their singleton form are banned from echo-word formation patterns (Müller 2004: 55). The rule is presented as a formula in Diagram 22.

$$[(C_{[\text{-bilabial\&nasal}]})V\text{-}]_{\text{reduplicand}} \rightarrow [mV\text{-}]_{\text{ECHO}}$$

Diagram 22: Rule for Turkish echo-word formation.

Depending on the presence or absence of a consonantal onset in the initial syllable of the reduplicand, the reduplication can thus be either complete or incomplete. Furthermore, the fixed segmentism of the echo always adds a modicum of inexactness. If the reduplication is incomplete, it is also always inexact. In the case of complete reduplication, the image contains an additional phonological unit which is absent from the reduplicand. The phonological correspondence of $A_{\text{reduplicand}}$ and A'_{image} is thus always inexact. Therefore, canonicity is impaired to some extent.

Moreover, the image is said to be "meaningless by itself" (Abbi 1992: 20). The echo-word construction, however, has a constructional meaning which is associated with vagueness, deprecation and/or indefinite plurality (Kallergi 2015: 18). There are thus already two factors which reduce the degree of canonicity because $A_{\text{reduplicand}}$ and A'_{image} are not fully identical phonologically and semantically and A'_{image} is not a syntactic word in isolation. This means that the criteria (C5), (C7), (C9), and (C10) of canonical reduplication are not met. If the characteristically fixed segmentism of the image is taken to have morpheme status of its own, the disagreement of echo-word formations and canonical reduplication would also include criterion (C8).

For the topic of this paper, it is especially important to note that it is possible to further differentiate echo-word constructions as to how non-canonical they are. The Ryukyuan languages of Okinawa make ample use of echo-word constructions (Otsuka 2016: 90–97). There are different types of echo-word constructions which differ from each other as to the number of segments of the reduplicand which undergo substitution in the image. There is the replacement of single consonants, syllable bodies, and entire syllables of the reduplicand as shown in Table 4 (based on Otsuka 2016: 92–97).[71] The CV-positions of the syllables which are relevant for the echo-word formation are presented in individual

71 On the basis of the information given in Otsuka (2016), it cannot be determined how vowel-initial reduplicands are copied in echo-word constructions. The phonological status of the glottal stop /ʔ/ is not entirely clear either so that words which display an initial phonetic [ʔ] might turn out to be phonologically vowel-initial.

cells. The remainder (= rhyming part) of the segmental chains is presented en bloc. Grey shading is indicative of those parts of the segmental chains which are identical for reduplicand and image. The fixed segments of the echo are highlighted in boldface.

Table 4: Types of echo-word constructions in Ryukyuan languages.

Type	Reduplicand					Image					Meaning
(A)	C	V				C	V				'strong congestion'
	ʧ	*i:*	*hai*			*n*	*i:*	*hai*			
(B)	C	V	C			C	V	C			'many difficulties'
	n	*a*	*n*	*ji*		***k***	***u***	*n*	*ji*		
(C)	C	V	C			C	V				'drop by repeatedly'
	t	*u*	*n*	*migui*		***k***	***e:***	*migui*			
(D)	C	V	C	V		C	V				'run in panic'
	ʔ	*a*	*ʃ*	*i*	*marubi*	***t***	***i:***	*marubi*			
(E)	C	V				C	V	**C**	**V**		'wrestling'
	b	*a:*	*ke:*			***k***	***a***	***r***	***a***	*ke:*	

Type (A) gives evidence of the replacement of the initial consonant of the reduplicand (/ʧ/ > /n/) whereas the syllable nucleus remains unaffected by the copying process. In the case of Type (B), the CV-body of the initial syllable of the reduplicand is replaced by another CV-body in the image (/na/ > /ku/). Thus, a sequence of two segments is involved in the replacement. Type (C) is an example of the replacement of an entire initial CVC-syllable with a CV-syllable (/tun/ > /ke:/), i.e., a sequence of three segments gives way to a combination of two segments. As to Type (D), the replacement ranges over two CV-syllables of the reduplicand whose substitute is a single CV-syllable in the image (/ʔaʃi/ > /ti:/). Thus, four segments are replaced with just two. The relation is reversed in Type (E). Here we find a CV-syllable of the reduplicand being replaced with a sequence of two CV-syllables in the image (/ba:/ > /kara/). This means that not only the number of segments has been doubled but also the number of syllables has increased by one.

The size of the phonological chains of $A_{reduplicand}$ and A'_{image} is the same for the first two types (A)–(B). In contrast, types (C), (D), and (E) show that the image may either count more or less segments than form part of the phonological chain of the reduplicand. If we put the focus on the parameter of segmental identity of $A_{reduplicand}$ and A'_{image} (= criterion (C7)), it is possible to put these types of echo-word constructions in an order of decreasing canonicity as shown in Diagram 23.

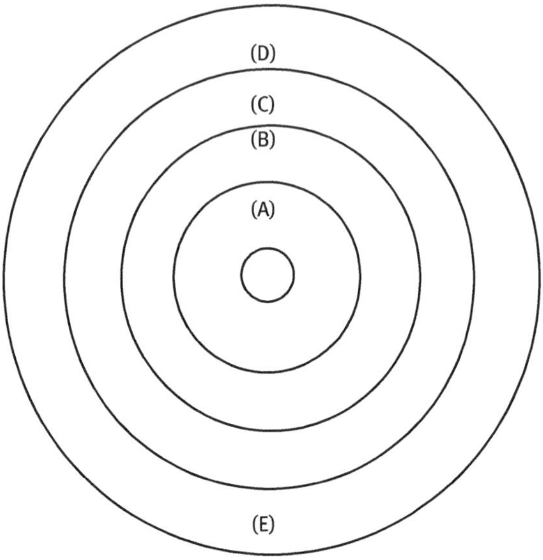

Diagram 23: Decreasing canonicity of Ryukyuan echo-word formations.

The circle at the center is left empty because none of the types of echo-word formation is identical to canonical reduplication. If it were, the notion of echo-word formation would become nonsensical.[72] The impossibility of a 100 % match of canon and echo-word notwithstanding, one cannot treat all types of echo-word formations as one in terms of their dissimilarity to canonical reduplication. To the contrary, if we look at their behavior under the parameter of phonological equivalence (= (C7)), a very clear pattern emerges according to which non-canonicity is a matter of degree and thus a gradable property. Type (A) is still relatively close to the canon whereas Types (D) and (E) are far removed from the center of Diagram 23.[73]

72 In the concluding Section 5, I discuss the possibility that echo-word formations (and other supposedly non-canonical types of reduplication) may have a canon of their own which obeys to principles which are different from those of proper reduplication.

73 I treat Types (D) and (E) as equidistant from the canon because there is as yet no satisfactory solution to the problem of differentiating phonologically reductive and phonologically augmentative cases of echo-word formation.

4.2.1.2 Cumulation of non-canonical properties

It is by no means rare cross-linguistically that echo-word formations escape being neatly classified according to patterns. This is the case for a variety of examples from Ryukyuan which Otsuka (2016: 100–101) terms exceptions. Similarly, Authier (2012: 44) in his grammar of Juhuri (Indo-European, Asia) states that certain loanwords are integrated into the system "en les redoublant avec une déformation assez aléatoire" as the noun *şele* 'burden' from Azerbaijani in (15):

(15) Juhuri [Authier 2012: 44]

vegi	***şele~şulte**=y=tü=re*	*bu=ra*	*ä=xune=y*	*bebe=şmu*
take	**burden~REDUP**=EZ=2=DAT	IMP=go	LOC=house=EZ	father=2PL

'Take **all of your stuff** and go back to your father's'

What reduplicand and image share phonologically are the disyllabicity, the initial fricative /ʃ/, the medial lateral /l/, and the final vowel /e/. The vowel of the first syllable is different (/e/ vs. /u/). Moreover, the image has an additional dental plosive /t/ and thus counts one segment in excess of the reduplicand (cf. Diagram 24).

A_{reduplicand}	*ş*	*e*	*l*		*e*
A'_{image}	*ş*	*u*	*l*	*t*	*e*

Diagram 24: An echo-word in Juhuri.

The grey shaded cells show that there is some similarity. Still, the segments which are shared by reduplicand and image do not form a continuous sequence because they are interrupted by phonological units which are unique to the reduplicand or the image. On account of this, it is questionable whether cases of this kind are acceptable as instances of reduplication generally because there is no systematicity that regulates the formation of these and other individual cases.

Echo-word constructions deviate from canonical reduplication on several parameters. To the violations of the canon mentioned in the previous Section 4.2.1.1, several additions can be made. Keane (2005) studies cases of phrasal reduplication as found in many languages of the Indian sub-continent. Phrasal reduplication can come in different shapes. In the bulk of the cases, the domain and the reduplicand are identical and comprise two syntactic words one of which is always a noun. In the example (16) from the Dravidian language Tamil, it is shown that phrasal reduplication is combinable with echo-word formation.

(16) Tamil [Keane 2005: 247][74]

avan	***nalla***	***paiyan***	***killa***	***paiyan**-ɲɲu*	*nampaatee*
he	**good**	**boy**	ECHO	**boy**-QUOTE	believe:NEG.IMPER

'Don't believe that he's **a good boy and so forth**!'

The phrasal nature of the construction is the first deviation from the canon because, according to criterion (C5), the latter expects $A_{reduplicand}$ and A'_{image} to be syntactic words – but not syntagms. In the case of (16), we have a head noun *paiyan* 'boy' which is preceded by the adjectival modifier *nalla* 'good'. This NP forms the input for the phrasal reduplication-cum-segmental modification. The output functions as the host of the quotative marker. Since we are dealing with an echo-word formation pattern at the same time, the parameters of formal identity of reduplicand and image are also violated against. In addition, the first constituent of the image starts with the CV-syllable /ki/ which replaces the initial syllable /na/ of the first constituent of the reduplicand so that the segmental chains of $A_{reduplicand}$ and A'_{image} differ from each other in two consecutive slots. Diagram 25 allows us to recapitulate the above observations.

Constituent	Adjective					Noun					
$A_{reduplicand}$	*n*	*a*	*l*	*l*	*a*	*p*	*a*	*i*	*y*	*a*	*n*
A'_{image}	*k*	*i*	*l*	*l*	*a*	*p*	*a*	*i*	*y*	*a*	*n*

Diagram 25: An echo-word in Tamil.

The Tamil case is particularly intriguing since it transcends the meaning-bearing unit which is supposed to define the upper limit of the size of the domain and the reduplicand. What adds to the problems posed by (16) and similar examples is the fact that the second constituent of the reduplicand undergoes reduplication without segmental modification so that it could pass as an instance of total reduplication.

Hungarian has a variety of types of so-called twin-words (Forgács 2007: 328).[75] There are patterns which are in line with what we know about echo-

74 Apart from deleting the square brackets which marked the boundaries of the phrasal reduplication in the original and glossing the second instance of *paiyan* as meaning 'boy', I have respected all conventions of my source.

words from languages like those discussed in the foregoing Section 4.2.1. In these cases, the initial consonant of the reduplicant is replaced with a labial consonant on the image as in *csonka~bonka* 'mutilated' with *csonka* (also with the meaning 'mutilated', cf. below) functioning as reduplicand. The initial affricate /ʧ/ (= <cs>) of the reduplicand is replaced with the voiced bilabial plosive /b/ on the image. If the reduplicand is vowel-initial, the image hosts a prosthetic labial plosive nevertheless as in *Andi* (the short form of the male first name András) → *Andi~Bandi* which has hypocoristic function. *Andi~Bandi* represents Type (I) whereas *csonka~bonka* is the representative of Type (J).

In the light of the above findings, there is nothing special about this class of Hungarian twin-words. The picture changes drastically, however, when we turn our attention to the apparently much larger class of cases which attest to the inverse linearization image > reduplicand. In Table 5, I include some of those cases which behave unexpectedly in the sense that the image precedes the reduplicand. I mark the segmental differences between image and reduplicand in bold on the image and by double underling on the reduplicand. The exact semantics of the so-called twin-words is not always clearly discernible. It is at times difficult to pinpoint the meaning difference between the reduplicand in isolation and the reduplicative construction.

Table 5: Twin-word constructions in Hungarian.

Type	Reduplicand		Twin-word construction		
	Form	Meaning	Image	Reduplicand	Meaning
(F)	*dombos*	'hilly'	*d**i**mb**e**s*	*dombos*	'hilly'
	görbe	'crooked'	*g**i**rbe*	*görbe*	'crooked'
(G)	*pici*	'small'	*ici*	*pici*	'very small'
	pirul	'to blush'	*irul*	*pirul*	'to blush'
(H)	*forog*	'to turn around'	***ireg***	*forog*	'to turn around'
	mozog	'to move'	***izeg***	*mozog*	'to wriggle'

Forgács (2007: 328) assumes three types of twin-words. As to (F), there is dissimilation in the sense that the vowels of the reduplicand and those of the image must belong to different classes according to the oppositions of front/back and labial/illabial. Type (G) comprises reduplicands with an initial labial consonant

75 All the Hungarian examples in the section are taken from Forgács (2007: 328–329).

which is dropped under copying. The result is an image that is by one segment shorter than the corresponding reduplicand. These are cases of apheresis. Since apheresis is tantamount to the loss of a segment, Type (G) is certainly not an instance of total reduplication. In the absence of fixed segmentism on the image, there is just the copy of a substring of the segmental chain of the reduplicand and thus, Type (G) has much more in common with partial reduplication. Very much the same can be said about Type (H) which combines the vowel-dissimilation rule of Type (F) with the apheresis rule of Type (G).

Diagram 26 reflects the case of *idres-fodros* 'ruffled and crinkled' based on *fodros* 'ruffled' ← *fodor* 'ruffle'. In contrast to the previous diagrams of this kind, the phonological chain of the image occupies the top line to match the inverted linearization of the constituents of the construction.

$A_{reduplicand}$		*i*	*d*	*r*	*e*	*s*
A'_{image}	*f*	*o*	*d*	*r*	*o*	*s*

Diagram 26: Hungarian twin-word of Type (H).

The situation is similar to that depicted in Diagram 24 for Juhuri. The phonological chains of reduplicand and image are only partly identical. Half of the segments of the reduplicand do not show up in the image. Even the size of the phonological chains is different. The Hungarian Type (H) is thus multiply non-canonical. The order image > reduplicand is not in line with criterion (C2). As in the previously discussed cases of echo-word formations, the requirements of criteria (C5) through (C10) are not met either. What is more, in many cases the criteria (C11) and (C13) are also not fulfilled since several of the twin-words seem to be full synonyms of the singleton item which functions as reduplicand.

I am now in a position to compare the degree of canonicity of echo-word constructions across languages. On the basis of the above evidence, I locate the different constructions from Turkish, Ryukyu, Juhuri, Tamil, and Hungarian on a radially organized hierarchy of decreasing canonicity in Diagram 27. The bracketed numbers refer to the sentential examples whereas the bracketed letters (A)–(E) represent Ryukyuan constructions and (F)–(J) those of Hungarian.

The different constructions are assigned their places on the different cycles on the basis of the number of criteria of canonical reduplication they violate against. In the case of an identical number of deviations, I further distinguish gradually between different degrees of dissimilarity of reduplicand and image on the phonological level by way of applying a very simple rule, namely the more segments reduplicand and image share the closer to the center the construction is placed and, the other way round, the more segments are different in a reduplicand-image pair the further away from the center the construction is placed. I acknowledge that this is a very crude procedure which calls for being refined in the future.

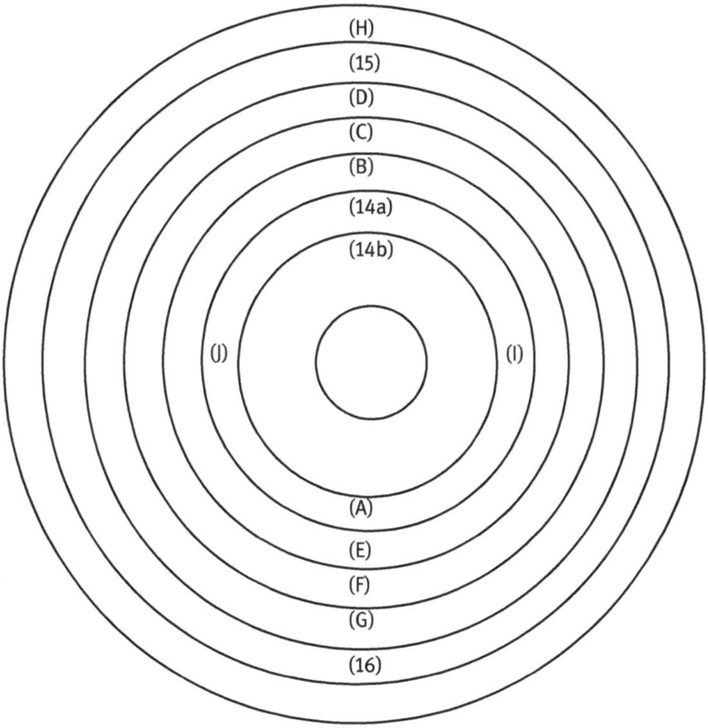

Diagram 27: Decreasing canonicity of echo-word constructions across languages.

Owing to the cumulation of several violations against the canon, the constructions (15), (16), and (H) wind up in the two outermost cycles. They are multiply non-canonical. The further we move to the inside of the radial category in Diagram 27, the closer we get to canonical reduplication. As the cases of (14b), (14a), (A), (I), and (J) show, even echo-word constructions can be relatively canonical. In general, the Hungarian twin-word constructions as of Table 5 do not fare well in terms of canonicity. In contrast, the Hungarian Types (I) and (J) which resemble the Turkish paradigm case of echo-word constructions are situated in the relative vicinity of the center of Diagram 27.

All five of the languages which are featured in this and the previous section also attest to total reduplication in the strict sense of the term. Table 6 contains one piece of evidence for each of the languages involved.

Table 6: Examples of total reduplication in languages with echo-word constructions.

Language	Singleton		Total reduplication			Source
	Reduplicand	Meaning	Reduplicand	Image	Meaning	
Turkish	*yer*	'place'	*yer*	*yer*	'at different places'	Müller (2004: 35)
Ryukyu	*itʃi-me:*	'one'[76]	*itʃi-me:*	*itʃi-me:*	'one each'	Otsuka (2016: 110)
Juhuri	*yeki*	'one'	*yeki*	*yeki*	'one each'	Authier (2012: 83)
Tamil	*(i)rənḍə*	'two'	*(i)rənḍə*	*(i)rənḍə*	'two each'	Abbi (1992: 79)
Hungarian	*öt*	'five'	*öt*	*öt*	'five each'	Forgács (2007: 327)

The constructions are bona fide cases of canonical reduplication. It is striking to see that all of these cases display distributive functions. The existence of canonical reduplication in the very same languages which give evidence of non-canonical reduplication in the shape of echo-words and the like proves that the central circles of the radial categories of the above Diagrams 23 and 27 are filled. Thus, for each of the above languages, the canon is not just an abstract ideal but a realized pattern which may serve as a language-internal reference point when the system of reduplication of a given language is inquired into.

The co-existence of canonical reduplication and echo-word formation does not exhaust the system of reduplication in each and every language. In point of fact, there is a multitude of phenomena which are at least as worthy of our interest as echo-words. For reasons of space, I can only discuss some of these equally intriguing phenomena in passing in the subsequent Section 4.3. To pay them the attention they deserve, I intend to elaborate on these issues in a follow-up study.

76 The Ryukyuan example is morphologically complex, viz. *itʃi-me:*= {NUM}-{CLASS}.

4.3 Beyond echo-words

4.3.1 Glimpses of partial reduplication

Partial reduplication has a plethora of realization forms which may differ from each other widely so that the study of partial reduplication alone is already a demanding task in terms of time and man-power. To keep the discussion within reasonable limits, I have to skip most of the still un(der)explored territory of partial reduplication.

I take example (17) from the Sinitic language Japhug as my starting point. What is of interest for the ensuing discussion is marked in bold in (17).

(17) Japhug [Jacques 2007: 17][77]

smɤn	***tɯ~ta**-ndza*	*nɯ*
medicine	RED-AOR.3SG-eat	NMLZ

'**all** the medicine s/he has taken'

Jacques (2007: 17) explains that "la réduplication sert à exprimer la totalité d'un ensemble d'action." Semantically, the Japhug case is thus in line with criterion (C13) of the canon. Formally, however, the degree of canonicity is minimal for several reasons. Since we are not dealing with two syntactic words which function as reduplicand and image the criteria (C5) through (C12) cannot be met in the first place. Additionally, the order of the reduplicand and the image runs counter to that imposed by the canon because we are dealing with prefixing reduplication, i.e. the image precedes the reduplicand so that also criterion (C2) is violated against. Furthermore, there is fixed segmentism which renders the image and the reduplicand dissimilar on the segmental level. The dissimilarity of image and reduplicand can be gathered from Diagram 28. The segments of the image occupy the top line to reflect the order image > reduplicand. The extra-bold line marks morpheme boundaries.

$A_{reduplicand}$	*t*	*ɯ*			
A'_{image}	*t*	*a*	*n*	*dz*	*a*

Diagram 28: Partial reduplication in Japhug.

77 I have turned the French glosses into English glosses. The English translation is based on the French translation of the example in my source.

The sole segment that is shared by reduplicand and image is the voiceless dental plosive /t/ in the syllable-onset position. The nucleus of the syllabic image is the obligatory high central vowel /ɯ/ which forms part of the image irrespective of the quality of the vowel in the reduplicand. The vowel /ɯ/ is a case of fixed segmentism (or of a duplifix in the parlance of Haspelmath & Sims (2010: 39)). Only the consonant immediately preceding the fixed segment can be considered a copy of a corresponding segment of the reduplicand. The reduplicand is always the initial syllable body of a verb (Jacques 2007: 10–14). If the verb already hosts a prefix (mostly tense/aspect or subject person markers), the syllable body of this prefix functions as reduplicand. In contrast to morpheme-based reduplication in Kwaza as illustrated by example (9), the Japhug case is based on syllable-phonological properties.

In Japhug, the onset position of syllables can be filled with up to three consonants so that beside singleton consonants CC- and CCC-clusters are admitted.[78] Under reduplication, these clusters behave differently. If the sonorants /w/, /l/, /r/, /j/ or the velar fricative /ɣ/ are part of a consonant cluster and occupy the position immediately before the vocalic nucleus, they are usually not copied unto the image if they are preceded by a consonant of a different phonological class. Consonants other than the above sonorants and the velar fricative are not affected by the simplification of clusters under reduplication.

In his account of Japhug reduplication, Jacques (2007: 11–12) assumes three types. On the information provided in my source I take the liberty to distinguish five types two of which are further divided in two subdivisons as shown in Table 7.

Table 7: Japhug patterns of partial reduplication.

Type	**Singleton**		**Partial reduplication**								
	Domain	Meaning	Image				Reduplicand				Rest
Ia	*mtɤm*	'he sees it'	*m*	*t*		*ɯ*	*m*	*t*		*ɤ*	*m*
Ib	*wzɟɯr*	'he changes it'	*w*	*z*	*ɟ*	*ɯ*	*w*	*z*	*ɟ*	*ɯ*	*r*
IIa	*pɣaʁ*	'he give it back'	*p*			*ɯ*	*p*	*ɣ*		*a*	*ʁ*
IIb	*zgroʁ*	'he attaches it'	*z*	*g*		*ɯ*	*z*	*g*	*r*	*o*	*ʁ*

78 My source does not mention cases of naked syllables which serve as input for the reduplicative process. Since Jacques (2007: 10) assumes that the nucleus and coda of the reduplicand are replaced with the fixed segment /ɯ/, syllables of the V(C)-kind would trigger the image ɯ- which leads to the identity of fixed segment and image. In the absence of more information on this issue, I can only speculate that the reduplication of vowel-initial syllables is blocked.

Type	Singleton		Partial reduplication								
	Domain	Meaning	Image				Reduplicand				Rest
III	*ɣrum*	‘it is white’	*ɣ*	*r*		*ɯ*	*ɣ*	*r*		*u*	*m*
IV	*lwoʁ*	‘he sprinkles it’	*l*			*ɯ*	*l*	*w*		*o*	*ʁ*
V	*fsroŋ*	‘he protects it’	*f*	*s*	*(r)*	*ɯ*	*f*	*s*	*r*	*o*	*ŋ*
			onset			FS	onset			V	C

The consonantal clusters of the reduplicand are completely copied unto the image in Types Ia, Ib, and III, whereas cluster simplification is a characteristic trait of Types IIa, IIb, and IV. Type V oscillates in the sense that it allows both preservation and simplification of the clusters. The size of the consonant cluster is not decisive for preservation or simplification to apply. As already mentioned, the crucial factor is the presence of one of the sonorants /w/, /l/, /r/, /j/ or the velar fricative /ɣ/ in the slot immediately to the left of the nucleus. None of these consonants is involved in Type Ia and Ib. They are relevant, however, for the remaining Types IIa–V. Cluster simplification is compulsory if the consonants under scrutiny are preceded by segments which rank lower on the sonority scale. This applies to Types IIa and IIb. If the preceding consonant is one of the set /w/, /l/, /r/, /j/, or /ɣ/, cluster simplification is barred. This is the basis of Type III. The situation is complicated by Type IV which comprises a number of putative exceptions to Type III. According to Jacques (2007: 12), cluster simplification takes place only if an approximant /w/ or /j/ is preceded by a liquid /l/ or /r/ in the reduplicand. In addition, /j/ and /r/ are deleted if they follow the sibilant /z/ or /ʑ/. Type V exemplifies optional cluster simplification which is probably connected to the relative neighborhood of fricatives and sonorants on the sonority scale.

The division into several types of reduplicative patterns is clearly phonologically induced. Nevertheless, it is possible to rank the above Types Ia–V according to the similarity of reduplicand and image. Diagram 29 is radially organized again with an empty central circle because of the absence of evidence of exact reduplication.

The presentation of the Japhug data only serves to illustrate that the evaluative model introduced in the previous sections is also applicable to partial reduplication in general. The results as shown in Diagram 29 are only fragmentary and provisional. I have only looked at the data from a single language and checked their behavior exclusively as to the segmental correspondence of reduplicand and image. It is clear that the analysis can only gain from including as many languages as possible with as many different types of partial reduplication as possible. Moreover an attempt must be made to integrate the hierarchies of partial reduplication into those of other types of (non-)canonical reduplication.

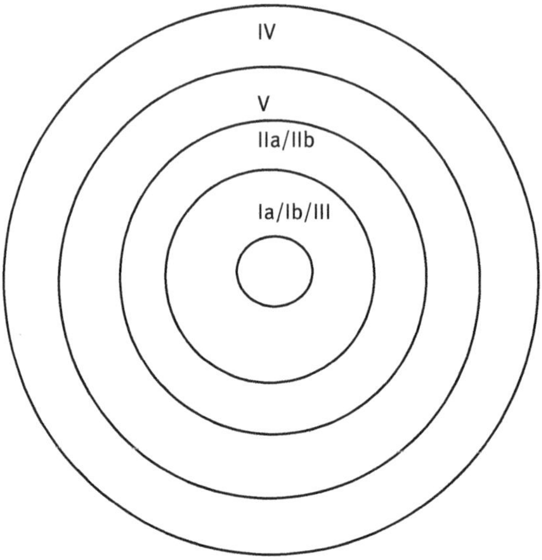

Diagram 29: Decreasing reduplicand-image similarity in Japhug.

4.3.2 Complicating things a bit further

Table 8 hosts examples of five additional phenomena which I propose to tackle summarily in this section. I discuss each of these phenomena only very briefly proceeding from top to bottom of Table 8.[79]

In her study of reduplication in Bikol, Mattes (2014: 34) states that "[s]yntactic repetition is considered a borderline case of reduplication by some authors." The Tagalog example in Table 8 does not only involve two identical syntactic words but also an additional element, the linker particle *na*, which joins the two instances of *pagod* 'tired'. The linker *na* functions very much like the coordinating conjunction *un* 'and' in the Estonian example (3).

79 This list is representative only of a selection of a huge variety of phenomena which are more or less closely related to canonical reduplication. For obvious reasons their discussion must be relegated to follow up studies.

Table 8: Further candidates for non-canonicity.

Phenomenon	Language/source	Example
syndetic repetition	Tagalog (Austronesian)/ Mattes (2014: 34)[80]	***Pagod*** *na* ***pagod*** *ako* **tired** LINK **tired** 1SG.AF 'I am **very tired**'
triplication	Mokilese (Austronesian)/ Harrison (1976: 223)[81]	***soal~soal~soal***-*da* *mijen* *loangge* **black~black~black**-DIR face:of sky:this 'The sky got **blacker and blacker**'
gemination	Pero (Chadic)/Al-Hassan (1998: 84)	SG *lóp* 'beat' → PL ***lóp~p*** 'beat'
empty reduplication	Tzeltal (Maya)/Mel'čuk (1996: 52)	SG *na* 'house' → AB.PL ***na~na***-*tik* 'lots of houses'
reduplicative stem-formation	Chamorro (Austronesian) / Stolz et al. (2009: 107)	*I* ***dengdeng*** *manhuyong* DEF **snail** PL:get_out *I* *liyang-ñiha* DEF cave-POR.3PL 'The **snails** emerged from their cave.'

As to the Tagalog case, Mattes (2014: 34) concedes that joining two identical syntactic words in this way is a common means of expressing intensification.[82] This means that the existence of a construction $[\text{X na X}]_{\text{intensive}}$ can be postulated. In this way the criteria (C11) through (C13) of canonical reduplication are met. The same holds for practically all other criteria from (C1) and (C2) via (C4) down to (C10). Therefore, one might ponder the idea that we are dealing with a phenomenon that is fully in line with the canon. However, there is the problem of the intercalated linker *na*. Owing to the presence of *na*, the two instances of *pagod* 'tired' are no direct neighbors on the syntagmatic axis so that the construction fails to fulfill the required adjacency according to criterion (C3). The linker characterizes the Tagalog case as an instance of syndesis, i.e. of a construction whose constituents are overtly connected to each other by additional morphological material. In a study dedicated to the languages of Europe, Stolz (2009: 108–111) shows that the distribution of functionally equivalent syndetic and asyndetic constructions over the languages of his sample reflects areal and genetic preferences. The patterns are such that, from a functionalist point of view, syndetic constructions cannot simp-

80 The example is originally from Schachter & Otanes (1972: 231).

81 I have added the morpheme glosses.

82 Tagalog also attests canonical reduplication with distributive numerals, for instance. Ramos (1971: 15) mentions cases like *isa* 'one' → *isa-isa* 'one by one'.

ly be excluded from a comprehensive stock-taking of reduplicative phenomena. Nevertheless, asyndesis is canonical whereas syndesis is non-canonical.

With reference to the example from Mokilese given in Table 8, Inkelas (2005: 70) argues that "triplication appears to be a requirement of the progressive construction when the input is monosyllabic", i.e. there are phonological conditions which trigger triplication. The presence of two images in lieu of one does not "correlate with a semantic change." To the mind of the quoted author, we are dealing with "semantically vacuous triplication." As a matter of fact, triplication is meaningless only in the sense that none of the images can be assigned a separate morpheme meaning. Harrison (1976: 224) compares several types of verbs in connection with their ability to triplicate. I reproduce his paradigms in Table 9.

Table 9: Reduplication and triplication in Mokilese.

Category	Semelfactive 'to walk'	Stative 'black'	Incorporation 'to drink'	Other 'to plant'
denotative	*kak*	*soal*	*nim pen*	*poadok*
progressive	*kak~kak*	–	*nim~nim pen*	*poad~poadok*
continuative	*kak~kak~kak*	*soal~soal~soal*	*nim~nim~nim pen*	*poad~poad~poadok*

What is crucial here is the empty cell for the progressive with stative verbs like *soal* 'black'. The absence of the progressive **soal~soal* from the paradigm of this stative verb is taken to suggest that it is the construction $[X\sigma\text{\textasciitilde}X\sigma\text{\textasciitilde}X\sigma]_{\text{continuative}}$ (with Xσ = Xσ) which prescribes that there must be three instances of monosyllabic verbs to form a grammatically acceptable construction. Alternatively, it is possible to assume that the input for the formation of the progressive of monosyllabic verbs is the singleton verb which then undergoes reduplication the output of which functions in turn as the input for the formation of the continuative. Under this analysis, the continuative is formed by reduplicating the initial CVC-syllable of the verb-form of the progressive. Semantically, the derivation of the continuative from the progressive is hardly a problem (Bybee et al. 1994: 172). Thus, triplication might not be a process of its own starting from the singleton verb but a succession of two consecutive processes of reduplication.[83] If this

83 As a solution for the problem posed by the Mokilese data, consecutive reduplication is not completely out of the way. Mattes (2014: 96–98) discusses the "[c]ombinations of various reduplication types" in one and the same word-form for Bikol. It is plausible that the combined reduplications reflect a chronology of steps with one reduplicative process providing the input

is the correct analysis, the necessity of considering triplication a distinct category diminishes considerably – at least as to the Mokilese case.[84] The empty cell is not a serious problem for the consecutive-reduplication analysis since analogy to verbs with a full paradigm of aspects can be invoked. My interpretation of the Mokilese data does not render the case fully canonical though because polysyllabic verbs of this language do not reduplicate totally so that the condition which requires reduplicand and image to be syntactic words is not obeyed. If, however, triplication is a distinct process criterion (C4) fails to apply which disallows the creation of multiple images.

Haspelmath & Sims (2010: 39) assume that reduplication "seems to have more in common with gemination or vowel lengthening." This statement can be interpreted the other way round in the sense that gemination and vowel lengthening have something in common with reduplication – and thus are potential members of the family of reduplicative processes. Similarly, El Zarka (2005: 388) argues in favor of viewing "gemination in Arabic verbs [...] as reduplication." Robbers (2016: 174–182) is more cautious in her reasoning on the basis of her sample of North American and Mesoamerican Indian languages. Nevertheless, she emphasizes that functionally consonant gemination and vowel lengthening cannot be categorically told apart from bona fide cases of reduplication. This is not different in the case of the Pero example in Table 8. The gemination of the stem-final consonant serves the purpose of expressing plurality.[85] On several of the parameters, gemination is clearly compatible with the criteria of canonicity, viz. (C1) through (C4) and (C11) through (C13). Non-canonical traits come to the fore in connection with the remainder of the criteria some of which are not applicable whereas others such as (C5) are clearly violated against.

---

for the subsequent process. Consider a word-form like *pig-**pu~puru**~pildit* {BG.UG}-**{IMPV}-{PL}**-{squeeze} 'squeezing' (Mattes 2014: 98) which involves the duplifix *Curu-* as plural marker. The initial *C-* of this duplifix copies the initial consonant of the lexical morpheme which happens to be /p/ so that the plural marker takes the shape *puru-*. This plural marker forms the domain of the next reduplication, namely that of the initial CV-body to yield the imperfective form of the verb. The initial CV-body of the plural marker is /pu/ which is copied as the image /pu/. The latter process is possible only after the verb has undergone pluralization.

84 According to Harrison (1976: 59–63), none of the different patterns of reduplication in Mokilese can be claimed for total reduplication uncontroversially. What superficially looks like total reduplication turns out to be the special behavior of monosyllables. However, there is also evidence of canonical reduplication in the domain of distributive numerals as e.g. with *emen* 'one' → *emen emen* 'one by one' (Harrison 1976: 108).

85 Canonical reduplication is also attested in Pero. Frajzyngier (1989: 143) provides an example with distributive-dispersive function, namely *ɗôk* 'one' → *ɗôk ɗôk* 'various ones'.

Reduplication is said to be empty if it only serves as an automatic (ancillary) process which goes along with say, affixation of the principal exponent of a given category. In the case of the plural of abundance in Tzeltal, the genuine marker of abundance is the suffix *-tik*.[86] The absence of meaning is at odds with the criteria (C10) through (C12) of the canon. As Stolz et al. (2011: 58–59) argue it strikes the eye that the supposedly empty reduplication is triggered in a context which belongs to the functional domain of reduplication, namely the expression of some kind of plurality.[87] The entire construction has a meaning that is absolutely in line with criterion (C13). This functional aspect notwithstanding, the Tzeltal case is still not fully canonical because it can be classified as an instance of root reduplication (cf. Section 4.1).

Reduplicative stem-formation is a cross-linguistically widely attested phenomenon. It differs from proper reduplication insofar as it is impossible to establish a reduplicand-image relationship. In the case of the Chamorro noun *dengdeng* 'snake', there are two phonologically identical closed syllables. However, the singleton **deng* is inexistant so that there is no reduplicand from which an image could be created in the first place. The syllable /deŋ/ carries no meaning and does not constitute a syntactic word. In sum, reduplicative stem-formation fails to meet the expectations as to criteria (C5) through (C13). It is therefore non-canonical to a very high degree.[88]

Discounting many details, I can summarize the above discussion as follows. The phenomena reviewed in this section are all underachievers in terms of canonicity. However, there are gradual differences in their non-canonical behavior. Many of the problems arise from the fact that $A_{reduplicand}$ and A'_{image} do not have the status of syntactic words. Some of the further shortcomings of theirs derive directly from the violation of criterion (C5). Closest to canonical reduplication we find syndetic repetition whose only deviation from the canon is connected to adjacency according to parameter (C3). All other phenomena scrutinized in this section are incompatible with more than one of the criteria of canonicity. Reduplicative stem-formation displays the highest number of violations of criteria so that it is located on the periphery of the radial category in Diagram 30.

---

86 Tzeltal also makes use of more canonical types of reduplication as in the formation of reciprocals like *mah* 'beat' → *mah~mah* 'beat each other' (Pfau & Steinbach 2005: 569, taken from Berlin 1963: 214). It is likely though that the monosyllabicity of the reduplicand is crucial so that the canonicity of the example must remain doubtful.

87 For a more detailed discussion of the problems posed by Tzeltal reduplication patterns, I refer the reader to Robbers (2016: 111–112 and 152).

88 In contrast to most of the languages reviewed in this paper, Chamorro lacks evidence of total reduplication completely (Stolz et al. 2015: 806–808).

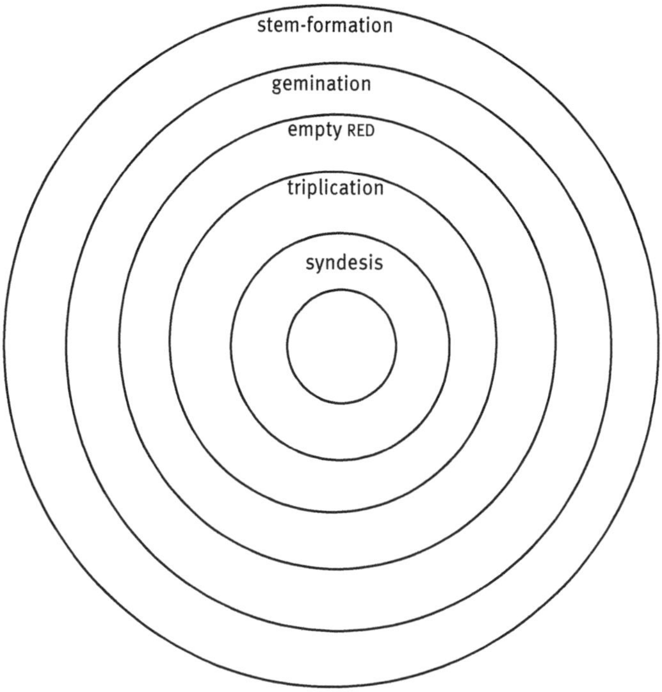

Diagram 30: Decreasing canonicity with selected phenomena.

The allocation of the phenomena on the radial category is based exclusively on the concrete examples discussed in the foregoing paragraphs. The present state of my knowledge does not allow us to generalize over all kinds of cases of syndesis, triplication, empty reduplication, and gemination, whereas I am confident that the peripheral position of reduplicative stem-formation can be proved to be robust across languages. Diagram 30 is not meant to be the final word on the issue of non-canonicity in the realm of reduplication. What it is meant to show nevertheless is the possibility of ordering reduplicative phenomena in a linguistically meaningful way.

5 Conclusions

The final sentence of the previous section can be taken to summarize very coarsely what this study has been intended to demonstrate, namely that the phenomenology of reduplication is not only immensely variegated but, at the same time, also equipped with an internal architecture. It is exactly this architecture that calls for being detected and further explored. To this end, a tool is needed which facilitates the comparison of different reduplicative constructions within individual languages as well as cross-linguistically. I have therefore suggested that the canonical approach is employed because it is independent from actually attested data and thus can be applied to any (sample of) language(s) without imposing the model of a given language as the universal reference point. In this spirit, I have put forward a first version of canonical reduplication to serve as the preliminary yard-stick in my search for the principles which underlie the architecture of reduplication language-independently. In this way, it has been possible to demonstrate that not only cases of putative reduplication but also bona fide instances of reduplication may differ widely as to the extent to which they correspond to or diverge from the canon. The variation in terms of the degree of canonicity is not arbitrary but reflects my choice of criteria which define canonical reduplication.

I am aware of the fact that the canon as presented in Sections 3.2–3.3 is a construct. This is unproblematic methodologically because I have explicitly looked for an abstraction of this kind to conduct my research without any empirically-induced bias. What I have presented in this paper is only a test run on the basis of a relatively unsophisticated and coarsely-grained prototype of canonical reduplication. There is no doubt that the canon is in urgent need of being revised, modified, and refined. The temporary nature of my proposal notwithstanding, the canon in its present form has already shown that it is very helpful when it comes to putting different reduplicative constructions in an order that makes sense linguistically. This does not mean that the problems associated with reduplication have already been settled in one go. I am still far from solving the riddles posed by the category under scrutiny. My study is only indicative of the direction to which future projects on reduplication should look.

Among the open questions which need to be answered urgently is that of the hierarchy of the criteria (C1)–(C13). In this paper, I have treated the thirteen criteria in an egalitarian way, i.e., all of them were given the same importance. However, it almost suggests itself that some of the criteria are more important than others. I therefore need to determine the relevance of each of the parameters for measuring canonicity. Some of the criteria are subordinated to first-order criteria. To reduce the number of criteria, it is possible to lump together the criteria (C6)–

(C10) because they relate to consequences triggered by criterion (C5). Moreover, neither is the list of parameters exhaustive as is nor is there a law of nature that requires certain criteria to be on the menu of canonicity in the first place.

I have given prominence to criterion (C5) by way of stipulating that the combination of two syntactic words in a reduplicative construction represents the canon. This decision of mine may be challenged if an alternative spell-out of the notion of reduplication is on offer. Several scenarios are conceivable among which there is the following which I find the most promising for triggering further advances in reduplication research. For the purpose of this paper, I have taken a very liberal approach in the sense that every kind of construction which has ever been labeled reduplication was entered into the data-base, in a manner of speaking. The continuum and the radial categories which I have presented in the previous sections seem to suggest that subsuming each and every supposedly reduplicative phenomenon under one and the same umbrella category is perhaps not the best option. From the position of the different constructions in the evaluative diagrams, it results that for several of the phenomena it is impossible – by definition – to reach the maximum score of canonicity. It can be asked therefore whether it would not make more sense to dissociate at least some of the types of reduplication from each other. The reason for this is the high probability that say, the systematic aspects of echo-word formation and perhaps also partial reduplication can be better captured if a distinct canon is assumed for each of these types. The postulation of different canons for total reduplication, partial reduplication, and echo-word formation is in line with the ideas of Hurch & Mattes (2005: 154) who assume that there is no (obligatory) grammaticalization cline that connects partial reduplication to total reduplication diachronically. My study has by no means proved that total reduplication, partial reduplication, and echo-word formation are phenomena which are largely autonomous from each other. The results are nevertheless suggestive of certain incompatibilities which distance the three types of reduplication from each other. It cannot be excluded that this is induced by my choice of canonical properties. Therefore, it is necessary to check whether the assumed dissociability of the three types of reduplication is corroborated also if an alternative canon is taken as the yard-stick. Much the same can be said in connection to the different types of total reduplication as discussed in Section 4.1. Do we have to assume that there are as many canons as there are types of total reduplication? A positive answer to this question might lead to the fragmentation of the category of reduplication into many smaller units – with the ultimate possibility of this very category disintegrating completely. To prevent this from happening we need to prove that one and the same canon can be applied to all of the competing types of total reduplication.

It could also be argued that I have started from the wrong point of departure because syntactic reduplication does not belong to the realm of reduplication in the first place because the word-boundary is not respected by the canonical constructions. From the point of view of functionalism, I rebut the exclusion of my canonical type of reduplication because syntactic reduplication can be shown to fulfill the same grammatical functions as word-based reduplication (Stolz et al. 2011: 148–157). Canonical reduplication as defined in this paper also allows us to take stock of the reduplicative construction types which compete with each other as to expressing identical functions. If however, an alternative canon yields results which demonstrate convincingly that syntactic reduplication has to be counted out to the benefit of my better understanding of reduplication in general, nothing will keep me from reconsidering my axioms.

Since this study forms but the first step towards developing the canon for reduplication, many tasks have to be tackled in the future. An attempt must be undertaken to integrate the above radial categories into one. It may turn out that it is relatively easy to measure canonicity and non-canonicity on particular parameters – especially criterion (C7) which involves counting segments – whereas it is relatively difficult to determine the gradual differences on other parameters. My prototype of canonical reduplication is thus far from perfection. In the near future, its viability has to be tested on a far richer empirical basis. Only then will it be feasible to evaluate the results typologically and in terms of linguistic theory. A possible hypothesis to work on in the future could assume that peripheral types of reduplication are attested less frequently across languages than those located near the center of the radial category. For the time being, however, I am content with inviting comments on my hypothesis that the canonical approach can be employed successfully also in the domain of reduplication research.

Acknowledgments: This study continues the research I have begun within the framework of the project STO 186/10-1 *Totale Reduplikation: Arealtypologie eines Universales / Total Reduplication: the Areal Typology of a Universal* financed by the *Deutsche Forschungsgemeinschaft* (DFG) in 2006–2007. Irrespective of the long time that has passed since then, I still want to express my gratitude to the DFG because it has become increasingly difficult in recent years to apply successfully for grants in my area of research. Julia Nintemann, Hitomi Otsuka, and Maja Robbers have kindly shared their expertise in the domain of the languages of the Americas, Africa, and Asia as to the properties of RED-constructions found in these areas. At the kitchen-sink, Christel Stolz kindly enlightened me on recursion as topic of the ongoing discussion about the indispensable prerequisites of human language. Nataliya Levkovych deserves to be mentioned specifically because of her kind help with the Modern Hebrew data and some of the technical aspects of this paper. I am grateful also to Cornelia Stroh for taking care of the editorial tasks connected to this paper. A word of thanks goes to Aina Urdze not only for lending me a helping hand with the Estonian data but also for giving me the opportunity to present an earlier version of this paper on occasion of the workshop on Non-prototypical reduplication (Bremen/Germany, 9–10 April, 2016). I am also grateful to my discussants at the very same workshop for their thought-provoking comments on the talk from which this paper developed. For all ideas – right or wrong – expressed in this study and the way in which they are argued, I assume the sole responsibility.

Abbreviations

1, 2, 3	first, second, third person
AB	abundance
ABL	ablative
ABS	absolutive
ADJ	adjective
ADV	adverb
AF	actor focus
AG	agentivizer
ALL	allative
AOR	aorist
ATT	attentive
AUX	auxiliary
BG	begun
C	consonant
CLASS	classifier
COMP	comparative
DAT	dative
DEC(L)	declarative
DEF	definite
DEM	demonstrative
DERIV	derivational
DET	determiner
DIR	directional
DISTRIB	distributive
ECHO	image in echo-word formation
ELA	elative
EMPH	emphatic
ERG	ergative
EXC	exclusive
EXI	existential
EZ	ezafe
F	feminine
FREQ	frequentative
FRUST	frustrative
FS	fixed segment
FUT	future
GEN	genitive
HES	hesitation
IMP(ER)	imperative
IMPV	imperfective
INF	infinitive
INFII	infinitive II
ITERV	iterative
LEX	lexical morpheme
LIM	limitative
LINK	linker
LOC	locative
N	noun
NEG	negative/negation
NMLZ	nominalizer
NOM	nominative
NUM	numeral
OBL	oblique
PL	plural
POR	possessor

PRES	present	TP	topic
PRET	preterit	TRSV	transitivizer
PROG	progressive	UG	undergoer focus
PROX	proximal	V	vowel
PST	past	φ	function/meaning
PTV	partitive	μ	morph(eme)
QUOTE	quotative	π	phonological (supra) segment
RECI	reciprocal		
RED(UP)	reduplication	σ	syllable
REFL	reflexive	ω	syntactic word
SG	singular	|	boundary of stem
SUPERESS	superessive	||	boundary of syntactic word
TNS	tense		

References

Abbi, Anvita. 1992. *Reduplication in South Asian Languages. An areal, typological and historical study*. New Dehli, etc.: Allied Publishers.

Aboh, Enoch O., Norval Smith & Anne Zribi-Hertz. 2012. Reduplication beyond the word level: A cross-linguistic view. In Enoch O. Aboh, Norval Smith & Anne Zribi-Hertz (eds.), *The morphosyntax of reiteration in Creole and non-Creole languages*, 1–26. Amsterdam/Philadelphia: Benjamins.

Al-Hassan, Bello S. Y. 1998. *Reduplication in the Chadic languages. A study of form and function*. Frankfurt a.M.: Lang.

Authier, Gilles. 2012. *Grammaire juhuri, ou judéo-tat, langue iranienne des Juifs du Caucase de l'est*. Wiesbaden: Reichert.

Bendel, Christiane. 2006. *Baskische Grammatik*. Hamburg: Buske.

Berlin, Brent. 1963. Some semantic features of reduplication in Tzeltal. *International Journal of American Linguistics* 29(3). 211–218.

Brandstetter, Renward. 1917. *Die Reduplikation in den indianischen, indonesischen und indogermanischen Sprachen*. Luzern: Kantonatsschule.

Bybee, Joan L. 2011. Markedness: Iconicity, economy, and frequency. In Jae Jung Song (ed.), *The Oxford handbook of linguistic typology*, 131–147. Oxford: Oxford University Press.

Bybee, Joan L., Revere Perkins & William Pagliuca. 1994. *The evolution of grammar. Tense, aspect, and modality in the languages of the world*. Chicago/London: The University of Chicago Press.

Corbett, Greville G. 2005. The canonical approach in typology. In Zygmunt Frajzyngier, Adam Hodges & David S. Rood (eds.), *Linguistic diversity and language theories*, 25–50. Amsterdam/Philadelphia: Benjamins.

Croft, William. 1993. Functional-typological theory in its historical and intellectual context. *Sprachtypologie und Universalienforschung* 46(1). 15–26.

Da Cruz, Aline. 2014. Reduplication in Nheengatu. In Gale Goodwin Gómez & Hein Van der Voort (eds.), *Reduplication in indigenous languages of South America*, 115–143. Leiden, Boston: Brill.

De Rijk, Rudolf P. G. 2008. *Standard Basque – a progressive grammar.* Cambridge, Mass/London: The MIT Press.

Dingemanse, Mark. 2015. Ideophones and reduplication: Depiction, description, and the interpretation of repeated talk in discourse. *Studies in Language* 39(4). 946–970.

Dressler, Wolfgang U. 1985. On the predictiveness of Natural Morphology. *Journal of Linguistics* 21. 321–337.

El Zarka, Dina. 2005. On the borderline of reduplication: Gemination and other consonant doubling in Arabic verbal morphology, In Bernhard Hurch (ed.), *Studies on reduplication*, 369–394. Berlin/New York: Mouton de Gruyter.

Erelt, Mati. 1997. Reduplication in Estonian. In Mati Erelt (ed.), *Estonian: Typological studies II*, 9–41. Tartu: University of Tartu, Department of Estonian.

Fabricius, Anne H. 1998. *A comparative survey of reduplication in Australian languages.* München/Newcastle: LINCOM Europa.

Fischer, Kerstin & Anatol Stefanowitsch. 2006. Konstruktionsgrammatik: ein Überblick. In Kerstin Fischer & Anatol Stefanowitsch (eds.), *Konstruktionsgrammatik. Von der Anwendung zur Theorie*, 2–18. Tübingen: Stauffenburg.

Forgács, Tamás. 2007. *Ungarische Grammatik.* Wien: Praesens.

Frajzyngier, Zygmunt. 1989. *A grammar of Pero.* Berlin: Reimer.

Goodwin Gómez, Gale & Hein Van der Voort. 2014. Reduplication in South America: An introduction. In Gale Goodwin Gómez & Hein Van der Voort (eds.), *Reduplication in indigenous languages of South America*, 1–16. Leiden, Boston: Brill.

Hannß, Katja & Pieter Muysken. 2014. Reduplication in Andean languages. In Gale Goodwin Gómez & Hein Van der Voort (eds.), *Reduplication in indigenous languages of South America*, 39–76. Leiden/Boston: Brill.

Harrison, Sheldon P. 1976. *Mokilese reference grammar.* Honolulu: The University Press of Hawaii.

Haspelmath, Martin. 1993. The diachronic externalization of inflection. *Journal of Linguistics* 42. 289–315.

Haspelmath, Martin. 2006. Against markedness (and what to replace it with). *Journal of Linguistics* 42(1). 25–70.

Haspelmath, Martin. 2008. Frequency vs iconicity in explaining grammatical asymmetries. *Cognitive Linguistics* 19(1). 1–33.

Haspelmath, Martin. 2011. The indeterminacy of word segmentation and the nature of morphology and syntax. *Folia Linguistica* 45(1). 31–80.

Haspelmath, Martin & Andrea D. Sims. 2010. *Understanding morphology.* London: Hodder Education.

Hurch, Bernhard & Veronika Mattes. 2005. Über die Entstehung von partieller Reduplikation. In Gertraud Fenk-Oczlon & Christian Winkler (eds.), *Sprache und Natürlichkeit. Gedenkband für Willi Mayerthaler*, 137–156. Tübingen: Narr.

Hurch, Bernhard & Veronika Mattes. 2007. The Graz database on reduplication. *Faits de Langues* 29. 191–202.

Hurch, Bernhard & Veronika Mattes. 2009. Typology of reduplication: The Graz database. In Martin Everaert, Simon Musgrave & Alexis Dimitriadis (eds.), *The use of data-bases in cross-linguistic studies,* 201–327. Berlin: Mouton de Gruyter.

Inkelas, Sharon. 2005. Morphological Doubling Theory. Evidence for morphological doubling in reduplication. In Bernhard Hurch (ed.), *Studies on reduplication*, 65–88. Berlin/New York: Mouton de Gruyter.

Inkelas, Sharon & Cheryl Zoll. 2005. *Reduplication. Doubling in morphology*. Cambridge: Cambridge University Press.

Jacques, Guillaume. 2007. La réduplication partielle en japhug, révélatrice des structures syllabiques. *Faits de langues* 29. 9–21.

Kager, René. 1999. *Optimality theory*. Cambridge: Cambridge University Press.

Kallergi, Haritini. 2015. *Reduplication at the word level. The Greek facts in typological perspective*. Berlin/Boston: De Gruyter Mouton.

Keane, Elinor. 2005. Phrasal reduplication and dual description. In Bernhard Hurch (ed.), *Studies on reduplication*, 239–261. Berlin/New York: Mouton de Gruyter.

Kouwenberg, Silvia & Darlene LaCharité. 2015. Arbitrariness and iconicity in total reduplication: Evidence from Caribbean Creoles. *Studies in Language* 39(4). 971–991.

Lakoff, George. 1987. *Women, fire, and dangerous things. What categories reveal about the mind*. Chicago/London: The University of Chicago Press.

Langemets, Margit, Mai Tiits, Tiia Valdre, Leidi Veskis, Ülle Viks, Piret Voll (eds.). 2009. *Eesti keele seletav sõnaraamat*. Tallinn: Eesti Keele Sihtasutus.

Lavotha, Ödön. 1973. *Kurzgefaßte estnische Grammatik*. Wiesbaden: Harrassowitz.

Levkovych, Nataliya. 2007. Totale Reduplikation im modernen Hebräischen. In Andreas Ammann & Aina Urdze (eds.), *Wiederholung, Parallelismus, Reduplikation. Strategien der multiplen Strukturanwendung*, 109–163. Bochum: Brockmeyer.

Lewis, Geoffrey L. 1969. *Turkish grammar*. Oxford: Clarendon Press.

Maas, Utz. 2005. Syntactic reduplication in Arabic. In Bernhard Hurch (ed.), *Studies on reduplication*, 395–430. Berlin/New York: Mouton de Gruyter.

Maas, Utz. 2007. Syntaktische Reduplikation: Stilistische Ressource vs. grammatische Form. Sprachbaudifferenzen in semitischen und indoeuropäischen Sprachen. In Andreas Ammann & Aina Urdze (eds.), *Wiederholung, Parallelismus, Reduplikation. Strategien der multiplen Strukturanwendung*, 1–28. Bochum: Brockmeyer.

Mattes, Veronika. 2014. *Types of reduplication. A case study of Bikol*. Berlin/Boston: De Gruyter Mouton.

Mel'čuk, Igor. 1996. *Cours de morphologie générale*. Vol. 3, Troisième partie: *Moyens morphologiques*; Quatrième partie: *Syntactiques morphologiques*. Montréal: Presse de l'Université de Montréal.

Miestamo, Matti, Dik Bakker & Antti Arppe. 2016. Sampling for variety. *Linguistic Typology* 20(2). 233–296.

Moravcsik, Edith. 1978. Reduplicative constructions. In Joseph H. Greenberg (ed.), *Universals of human language*. Vol. 3: *Word structure*, 297–334. Stanford: Stanford University Press.

Müller, Hans-Georg. 2004. *Reduplikationen im Türkischen. Morphophonologische Untersuchungen*. Wiesbaden: Harrassowitz.

Neukom, Lukas & Manideepa Patnaik. 2003. *A grammar of Oriya*. Zürich: Universität Zürich, Seminar für Allgemeine Sprachwissenschaft.

Nintemann, Julia. 2016. A comparative analysis of forms and functions of reduplication in Kenyan languages. In Christel Stolz & Thomas Stolz (eds.), *From Africa via the Americas to Iceland. Studies on reduplication and experiencers*, 1–93. Bochum: Brockmeyer.

Nintemann, Julia. this volume. Circumventing bisyllabic minimality. How to reduplicate monosyllabic verbs in (South-)East African Bantu languages.

Njika, Justina Atemajong. 2012. Nweh reduplication patterns. In Pius N. Tamanji & Gabriel M. Mba (eds.), *Aspects of reduplication in languages of Cameroon and Senegal*, 103–118. München/ Newcastle: LINCOM Europa.

Otsuka, Hitomi. 2016. *Reduplikation in den Sprachen Japans (und benachbarter Regionen).* Bochum: Brockmeyer.
Pfau, Roland & Markus Steinbach. 2005. Backward and sideward reduplication in German Sign Language. In Bernhard Hurch (ed.), *Studies on reduplication*, 569–594. Berlin/New York: Mouton de Gruyter.
Pott, August Friedrich. 1862. *Doppelung (Reduplikation, Gemination) als eines der wichtigsten Bildungsmittel der Sprache, beleuchtet aus Sprachen aller Welttheile.* Lemgo/Detmold: Meyer'sche Hofbuchhandlung.
Ramos, Teresita V. 1971. *Tagalog structures*. Honolulu: University of Hawai'i Press.
Robbers, Maja. 2016. Total, partial, neither, or both? A comparative study of reduplication in the languages of North and Central America. In Christel Stolz & Thomas Stolz (eds.), *From Africa via the Americas to Iceland. Studies on reduplication and experiencers*, 95–213. Bochum: Brockmeyer.
Rozhanskiy, Fedor Ivanovich. 2015. Two semantic patterns of reduplication: iconicity revisited. *Studies in Language* 39(4). 992–1018.
Rubino, Carl. 2005a. Reduplication. In Martin Haspelmath, Matthew Dryer, David Gil & Bernard Comrie (eds.), *The world atlas of language structure*, 114–117. Oxford: Oxford University Press.
Rubino, Carl. 2005b. Reduplication: Form, function and distribution. In Bernhard Hurch (ed.), *Studies on reduplication*, 11–30. Berlin/New York: Mouton de Gruyter.
Schachter, Paul & Fe T. Otanes. 1972. *Tagalog reference grammar.* Berkeley/Los Angeles/ London: University of California Press.
Southern, Mark R. V. 2005. *Contagious couplings. Transmission of expressives in Yiddish echo-phrases.* Westport,Connecticut/London: Praeger.
Stefanowitsch, Anatol & Stefan Th. Gries. 2005. Covarying collexemes. *Corpus Linguistics and Linguistic Theory* 1. 1–46.
Steriade, Donca. 1988. Reduplication and syllable transfer in Sanskrit and elsewhere. *Phonology Yearbook* 5(1). 73–155.
Stolz, Thomas. 2006. (Wort-)Iteration: (k)eine universelle Konstruktion. In Kerstin Fischer & Anatol Stefanowitsch (eds.), *Konstruktionsgrammatik. Von der Anwendung zur Theorie*, 105–132. Tübingen: Stauffenburg.
Stolz, Thomas. 2007a. Das ist doch keine Reduplikation! Über falsche Freunde bei der Suche nach richtigen Beispielen. In Andreas Ammann & Aina Urdze (eds.), *Wiederholung, Parallelismus, Reduplikation. Strategien der multiplen Strukturanwendung*, 47–80. Bochum: Brockmeyer.
Stolz, Thomas. 2007b. Re: duplication. Iconic vs. counter-iconic principles (and their areal correlates). In Paolo Ramat & Elisa Roma (eds.), *Europe and the Mediterranean as linguistic areas. Convergencies from a historical and typological perspective*, 317–350. Amsterdam/Philadalphia: Benjamins.
Stolz, Thomas. 2009. Total reduplication: Syndetic vs. asyndetic patterns in Europe. *Grazer Linguistische Studien* 71. 99–114.
Stolz, Thomas & Nataliya Levkovych. accepted. Function vs. form – on ways of telling repetition and reduplication apart. In Ulrike Feywald & Rita Finkbeiner (eds.), *Exact repetition in grammar and discourse.* Berlin/Boston: De Gruyter Mouton.
Stolz, Thomas, Nataliya Levkovych & Barbara Dewein. 2009. Reduplicative stem-formation: A comparative look at Maltese and Chamorro. *Ilsienna – Our Language. Working Papers of the International Assoiciation of Maltese Linguistics* 1. 87–133.

Stolz, Thomas, Cornelia Stroh & Aina Urdze. 2011. *Total reduplication. The areal linguistics of a potential universal.* Berlin: Akademie Verlag.

Stolz, Thomas, Aina Urdze, Julia Nintemann & Marina Tsareva. 2015. When some dots turn a different color ...: Thoughts on how (not) to determine whether or not reduplication is universal. *Studies in Language* 39(4). 795–834.

Van der Voort, Hein. 2014. Is reduplication an areal feature of the Guaporé-Mamoré region? In Gale Goodwin Gómez & Hein Van der Voort (eds.), *Reduplication in indigenous languages of South America*, 427–462. Leiden/Boston: Brill.

Viitso, Tiit-Rein. 2003. Structure of the Estonian language. Phonology, morphology and word formation. In Mati Erelt (ed.), *Estonian language*, 9–92. Tallinn: Estonian Academy Publishers.

Wälchli, Bernhard. 2005. *Co-compounds and natural coordination*. Oxford: Oxford University Press.

Wälchli, Bernhard. 2007. Ko-Komposita (im Vergleich mit Parallelismus und Reduplikation). In Andreas Ammann & Aina Urdze (eds.), *Wiederholung, Parallelismus, Reduplikation. Strategien der multiplen Strukturanwendung*, 81–108. Bochum: Brockmeyer.

Webster's. 1994. *Webster's Encyclopedic Unabridged Dictionary of the English Language*. New York: Gramercy Books.

Appendix

[Languages referred to in this paper]

Basque (Isolate, Europe)
Bikol (Austronesian, Asia)
Chamorro (Austronesian, Oceania)
Estonian (Uralic, Europe)
Hebrew, Modern (Afroasiatic, Asia)
Hungarian (Uralic, Europe)
Japhug (Sinitic, Asia)
Juhuri (Indo-European, Asia)
Kriol (English-based Creole, Australia)
Kwaza (Isolate, America)
Luiseño (Uto-Aztecan, America)
Mokilese (Austronesian, Oceania)
Nez Perce (Sahaptian, America)
Nheengatu (Tupi, America)
Nweh (Bantu, Africa)
Oriya (Indo-European, Asia)
Pero (Chadic, Africa)
Ryukyu (Japanese-Ryukyu, Asia)
Siwu (Kwa, Africa)
Tagalog (Austronesian, Asia)
Tamil (Dravidian, Asia)
Tarma Quechua (Andean, America)
Tausug (Austronesian, Asia)
Turkish (Turkic, Europe/Asia)
Tzeltal (Mayan, America)

Index of Authors

https://doi.org/10.1515/9783110599329-008

Index of Languages

https://doi.org/10.1515/9783110599329-009

Index of Subjects

https://doi.org/10.1515/9783110599329-010

Bei Fragen zur Produktsicherheit wenden Sie sich bitte an:
If you have any questions regarding product safety,
please contact:

Walter de Gruyter GmbH
Genthiner Straße 13
10785 Berlin
productsafety@degruyterbrill.com